Engineering Tools for Corrosion

Series Introduction

The EFC, founded in 1955, is a federation of 33 societies with interests in corrosion, and is based in 26 different countries throughout Europe and beyond. Its member societies represent the corrosion interests of more than 25,000 engineers, scientists, and technicians. The Federation's aim is to advance the science of the corrosion and protection of materials by promoting cooperation in Europe and collaboration internationally. Aside from national and international corrosion societies, universities, and research centers, companies can also become affiliate members of the EFC.

The administration of the Federation is in the hands of the Board of Administrators, chaired by the EFC president. The scientific and technical affairs are the responsibility of the Science and Technology Advisory Committee, chaired by the STAC chairman, assisted by the scientific secretary. The General Assembly approves any EFC policy prepared and presented by the BoA.

The Federation is managed through its General Secretariat, with three shared headquarters located in London, Paris and Frankfurt.

The EFC carries out its most important activities through its 19 active working parties devoted to various aspects of corrosion and its prevention, covering a large range of topics: corrosion and scale inhibition, corrosion by hot gases and combustion products, nuclear corrosion, environment sensitive fracture, surface science and mechanisms of corrosion and protection, physicochemical methods of corrosion testing, corrosion education, marine corrosion, microbial corrosion, corrosion of steel in concrete, corrosion in oil and gas production, coatings, corrosion in the refinery industry, cathodic protection, automotive corrosion, tribo-corrosion, corrosion of polymer materials, corrosion protection of drinking water systems, corrosion of archaeological and historical artifacts. The EFC is always open to formulating new working parties in response to the demands brought about by developing technologies, and their ensuing corrosion requirements and applications.

The European Federation of Corrosion's flagship event is EUROCORR, the most important corrosion congress in Europe, which is held annually in a different European country in September of each year. To date, 28 EUROCORR conferences have taken place in 12 different countries, and they have gained a reputation for their high technical quality, global perspective, and enjoyable social program. Another channel for the EFC's valuable transfer of knowledge is the EFC "green" book series, which is the fruit of the collaboration and high scientific caliber within and amongst the EFC working party members, and is emblematic of the EFC editorial policy.

In 2012, the EFC concluded an agreement with Woodhead Publishing Limited to publish new titles in the EFC's prestigious book series, drawing together research from the various EFC working parties. The Woodhead Publishing is now an imprint of Elsevier S&T Books. EFC Offices are located at:

European Federation of Corrosion, The Institute of Materials, Minerals and Mining, 1 Carlton House Terrace, London SWIY 5DB, UK

Fédération Européenne de la Corrosion, Fédération Française pour les sciences de la Chimie, 28 rue Saint-Dominique, F-75007 Paris, France

Europäische Föderation Korrosion, DECHEMA e.V., Theodor-Heuss- Allee 25, D-60486 Frankfurt-am-Main, Germany

European Federation of Corrosion Publications
Number 68

Engineering Tools for Corrosion

Design and Diagnosis

Luciano Lazzari

Published for the European Federation of Corrosion
by Woodhead Publishing

ELSEVIER

WP
WOODHEAD
PUBLISHING
An imprint of Elsevier

EFC
EUROPÄISCHE FÖDERATION KORROSION
EUROPEAN FEDERATION OF CORROSION
FÉDÉRATION EUROPÉENNE DE LA CORROSION

Published for the European Federation of Corrosion by Woodhead Publishing, an imprint of Elsevier S&T Books

Woodhead Publishing is an imprint of Elsevier
The Officers' Mess Business Centre, Royston Road, Duxford, CB22 4QH, United Kingdom
50 Hampshire Street, 5th Floor, Cambridge, MA 02139, United States
The Boulevard, Langford Lane, Kidlington, OX5 1GB, United Kingdom

British Library Cataloguing-in-Publication Data
A catalogue record for this book is available from the British Library

Library of Congress Cataloging-in-Publication Data
A catalog record for this book is available from the Library of Congress

ISBN: 978-0-08-102424-9 (print)
ISBN: 978-0-08-102425-6 (online)

For information on all Woodhead Publishing publications
visit our website at https://www.elsevier.com/books-and-journals

Publisher: Mathew Deans
Acquisition Editor: Christina Gifford
Editorial Project Manager: Andrea Gallego Ortiz
Production Project Manager: Surya Narayanan Jayachandran
Cover Designer: Greg Harris

Typeset by MPS Limited, Chennai, India

Dedication

This book is dedicated to the memory of my mentors
Pietro Pedeferri (1938–2008) and
Giuseppe Bianchi (1919–1996)

Contents

Author Biography

Luciano Lazzari graduated at the Faculty of Chemical Engineering of Politecnico di Milano, Italy, with a scientific investigation in corrosion in 1972. He was researcher in the field of corrosion from 1973 to 1978. He moved to AGIP SpA, State Italian Oil Company, now ENI SpA, as corrosion engineer until 1985. Since 1985 he has been a teacher in Corrosion and Protection at the Politecnico di Milano, associate professor from 1998 and full professor from 2005 in Material Science and Engineering. Since 1986 he has dealt with professional activity in the field of corrosion and prevention, in the Italian company Cescor srl of which he is President, with particular emphasis on selection of materials in design phase and cathodic protection. He has written a large number of articles in corrosion and cathodic protection; he is co-author of the book "Cathodic Protection" with Prof. P. Pedeferri. He is a member of NACE and EFC. He was co-chairman of the European Corrosion Congress in 2014 at Pisa, Italy (Eurocorr 2014). Presently, he retired from Politecnico di Milano and continues consultancy activities at Cescor.

List of Symbols and Abbreviations

α	Scale parameter (Gumbel distribution)
$a_{M^{z+}}$	Activity (or concentration) of ions of metal M in a solution (mol/L)
β	Weibull module (shape parameter)
b, b_a, b_c	Tafel slope (module) (V/decade)
BEM	Boundary Element Method
CA	Corrosion Allowance (mm)
C_{rate}	Corrosion rate (mm/year; μm/year)
$C_{rate,atm}$	Corrosion rate in corrosive atmospheres (μm/year)
$C_{rate,DA}$	Corrosion rate in Differential Aeration (mm/year)
$C_{rate,GC}$	Corrosion rate in Galvanic Corrosion (mm/year)
$C_{rate,IC}$	Corrosion rate in Intergranular Corrosion (mm/year)
$C_{0,atm}$	Initial corrosion rate in corrosive atmospheres (μm/year)
$C_{rate,SRB}$	Corrosion rate in SRB attack (mm/year)
CCC	Critical Chloride Concentration (g/L or ppm)
CCGS	Critical Crevice Gap Size (μm)
CIPP	Close Interval Potential Profile
CIT	Crevice Induction Time (h)
CP	Cathodic Protection
CSE	Copper sulphate electrode (+0.30 V SHE)
CV	Coefficient of variation
c.d.	Current density
D	Diffusion coefficient (m^2/s)
d	Distance (m)
δ	Scale parameter (Weibull distribution)
δ	Diffusion-layer thickness (m)
DL, DL_{min}	Design life, Minimum Design Life (years)
ΔE	Driving voltage or potential difference (V)
ΔM	Mass loss (kg)
e^-	Electron
EMF	Electromotive force (V)
E	Electrode potential (V)
E^0	Standard potential (V)
$E_{eq,A}$	Equilibrium potential of anodic process (V)
$E_{eq,C}$	Equilibrium potential of cathodic process (V)
E_A	Anodic potential (V)
E_C	Cathodic potential (V)
E_{Cl_2}	Potential of chlorine reduction cathodic reaction (V)
E_{corr}	Free-corrosion potential (V)
$E_{corr,A}$; $E_{corr,C}$	Free-corrosion potential of metals in a galvanic coupling (V)
E_{O_2}	Potential of oxygen reduction cathodic reaction (V)

$E_{O_2/bio}$	Potential of oxygen reduction cathodic reaction in the presence of biofilm (V)
E_{pit}	Critical pitting potential or passivity breakdown potential (V)
E_p	Passivation potential (V)
E_{pp}	Primary passivation potential (V)
E_{prot}	Protection potential (V)
E_{rp}	Repassivation potential (V)
E_{tr}	Transpassivity potential (V)
η	Overvoltage with respect to the equilibrium potential (V)
η_A	Anodic overvoltage (V)
η_{att,O_2}	Overvoltage activation of oxygen reduction (V)
η_C	Cathodic overvoltage (V)
η_{conc,O_2}	Overvoltage concentration of oxygen reduction (V)
η_H	Overvoltage activation of hydrogen evolution reaction (V)
η_M	Overvoltage activation of metal dissolution reaction (V)
η_{O_2}	Overvoltage of oxygen reduction (V)
f	Confidence for median rank
F	Faraday constant (26.8 A · h = 96,500 C = 23,060 cal/V · eq)
F	Confidence in median rank
FEA	Finite Element Analysis
FEM	Finite Element Method
$F_{conf,m,n}$	Fisher Distribution Function
$F(i)$	Median Rank Function
$F(t)$	Failure function, cumulative of f(t)
$f(t)$	Probability density function (pdf)
ϕ	diameter of a tube, ϕ_{tube} or a pipe, ϕ_{pipe} or an anode ϕ_{anode} (m)
$\phi(z)$	Hasting equation for normalized normal distribution
γ	Threshold parameter (Weibull distribution)
γ	Density (Mg/m^3)
Γ	Gamma function
G	Gibbs free energy (J/mol)
ΔG°	Standard Gibbs free energy variation (J/mol)
H	Activation energy (J/mol)
ICCP	Impressed Current Cathodic Protection
I	Current (A)
I_a	Anodic current (A)
I_c	Cathodic current (A)
I_e	External current (A)
I_{el}	Current in the electrolyte (A)
I_{prot}	Protection current (A)
i	Current density (mA/m^2)
i_a, i_c	Anodic, cathodic current density (mA/m^2)
i_{corr}	Corrosion current density, Corrosion rate (mA/m^2)
$i_{corr,acid}$	Corrosion rate in acids (A/m^2)
$i_{corr,strong\text{-}acid}$	Corrosion rate in acids (A/m^2)
i_{corr,CO_2}	Corrosion rate in carbonic acid (A/m^2)
i_{cp}	Critical passivation current density (A/m^2)
i_L	Oxygen limiting current density (mA/m^2)
i_0	Exchange current density (mA/m^2)
$i_{0,H}$	Exchange current density of hydrogen (mA/m^2)

$i_{0,M}$	Exchange current density of metal M (mA/m^2)
$i_{0,O}$	Exchange current density of oxygen (mA/m^2)
i_p	Passivity current density (mA/m^2)
i_{prot}	Protection current density (mA/m^2)
IR_{el}	Ohmic drop in the electrolyte (V)
k	constant (generic)
k_M	erosion–corrosion constant
K_s	Complex stability constant
L	Length (m)
L_{max}	Throwing power (m)
log	Logarithm base 10
ln	Natural logarithm (base *e*)
LSI	Langelier Saturation Index
λ	Location parameter (Gumbel distribution)
m	Mass (g, kg)
M	Generic metal
M^{z+}	Oxidized species of a generic metal
MIC	Microbiological Induced Corrosion
MOB	Manganese Oxidizing Bacteria
MTTF	Mean time to failure (h; years)
MTL	Maximum thickness loss (μm; mm)
M_W	Atomic or molecular mass (g/mol)
μ	mean (expected value)
N	Number of samples
ν	Cinematic viscosity of a fluid (poise)
P	Expected probability (in sampling)
P	Pressure of a gas (bar)
pCO_2	Partial pressure of CO_2 (bar)
p	Porosity of a scale (unitary fraction)
p	Prevalence (unitary fraction)
pdf	Probability density function or $f(x)$
P_{exp}	Expected confidence (in sampling)
PIT	Pitting Induction Time (h)
PREN	Pitting Resistance Equivalent Number
Q	Flux of electrical charges (A/m^2)
ψ_{oh}	Ohmic drop in an electrolyte (V)
r	Ratio of parameters in electrochemical modelling
ρ	Resistivity ($\Omega \cdot m$)
ρ_{el}	Electrolyte resistivity ($\Omega \cdot m$)
R	Resistance (generic ohmic resistance) (Ω)
R	Gas constant ($1.987\ cal \cdot k \cdot eq = 8.314\ J/mol \cdot K$)
Re	Reynold number (adimensional)
RH	Relative humidity
$R(t)$	Reliability function
R_{tot}	Total resistance (Ω)
σ	standard deviation
s	Thickness (mm or μm)
Δs	Thickness loss (mm or μm)
S; S_A; S_C	Surface; anodic surface; cathodic surface (m^2)

SCC	Stress Corrosion Cracking
Sc	Schmidt number (adimensional)
Sh	Sherwood number (adimensional)
SHE	Standard Hydrogen Electrode
SCE	Saturated Calomel Electrode (+0.24 V SHE)
SRB	Sulphate Reducing Bacteria
s.v.	Stochastic variable
t	Time (s; h; d; years)
t_{PIT}	Pitting Induction Time (h)
T	Temperature (°C; K)
T	Return period
TDS	Total Dissolved Solids or Salinity (g/L or mg/L)
TTF	Time to failure
v	Velocity (m/s)
v_{E-C}	Erosion–Corrosion Critical Velocity (m/s)
V	Voltage (V)
ΔV	Voltage drop or ohmic drop (V)
WF	Wetting Factor
x_{max}	Extreme value (Gumbel distribution)
Y	Generic corrosion affecting parameter
z	Valence, number of electrons in an electrochemical reaction
z	Normalized variable

Units

A	Ampere
A · h	Ampere hour
C	Coulomb
°C	Degree centigrade
h	hour
J	Joule
K	Kelvin
kWh	Kilowatt hour (1 kWh = 800 kcal = 3.6×10^3 kJ)
L	Litre
m	Metre
mol	Mole
Ω	Ohm
s	Second
S	Siemens
V	Volt

Premise

This book has a twofold aim to:

- improve the use of an engineering approach in design, operating and diagnosis;
- suggest models, algorithms and empirical equations for estimation of main relevant parameters such as expected corrosion rate, electrochemical conditions, thresholds limits, initiation time.

Models and equations here proposed are derived from the theory, validated by the experience and by computer modelling when dealing with electrical field. Most of the simplified algorithms proposed has to be considered the result of the activities carried out by the, as I like to call, *Pietro Pedeferri's school on corrosion* at *Politecnico di Milano*, Milan, Italy, since 1960s and continued by PoliLaPP laboratory. I have also a debt with prof. *Giuseppe Bianchi* who taught corrosion at *The State University of Milan*, Milan, Italy, for almost five decades: from them I learned the importance of the conjugation of the theory and the engineering practice.

Beyond the data gathered from experience and laboratory investigations, the central thread of this approach has been the belief that the theory beneath the corrosion science as well as the corrosion engineering allows to gain models and tools for sound calculation and reliable predictions. Here, experimentation and literature data have been used for validation of models and algorithms rather than to derive them empirically.

In this book, basic theory is missed, since I considered that users and readers already have a robust knowledge of it. Reference has to be made to classical handbooks on corrosion science and corrosion engineering as well as the new *Pedeferri's Corrosion and Protection of Metals*.

Finally, I warmly thank all my colleagues and workmates at PoliLaPP and Cescor, with whom I discussed the algorithms, the models and their applications. Moreover, my hope is that young corrosion engineers will use the algorithms I propose here and perhaps try improving either their formulations or the empirical constants.

Luciano Lazzari
Milano, March 2017

Volumes in the EFC Series List

1. **Corrosion in the nuclear industry**
   Prepared by Working Party 4 on Nuclear Corrosion
2. **Practical corrosion principles**
   Prepared by Working Party 7 on Corrosion Education (out of print)
3. **General guidelines for corrosion testing of materials for marine applications**
   Prepared by Working Party 9 on Marine Corrosion
4. **Guidelines on electrochemical corrosion measurements**
   Prepared by Working Party 8 on Physico-Chemical Methods of Corrosion Testing
5. **Illustrated case histories of marine corrosion**
   Prepared by Working Party 9 on Marine Corrosion
6. **Corrosion education manual**
   Prepared by Working Party 7 on Corrosion Education
7. **Corrosion problems related to nuclear waste disposal**
   Prepared by Working Party 4 on Nuclear Corrosion
8. **Microbial corrosion**
   Prepared by Working Party 10 on Microbial Corrosion
9. **Microbiological degradation of materials and methods of protection**
   Prepared by Working Party 10 on Microbial Corrosion
10. **Marine corrosion of stainless steels: chlorination and microbial effects**
    Prepared by Working Party 9 on Marine Corrosion
11. **Corrosion inhibitors**
    Prepared by the Working Party on Inhibitors (out of print)
12. **Modifications of passive films**
    Prepared by Working Party 6 on Surface Science
13. **Predicting CO_2 corrosion in the oil and gas industry**
    Prepared by Working Party 13 on Corrosion in Oil and Gas Production (out of print)
14. **Guidelines for methods of testing and research in high temperature corrosion**
    Prepared by Working Party 3 on Corrosion by Hot Gases and Combustion Products
15. **Microbial corrosion: Proceedings of the 3rd International EFC Workshop**
    Prepared by Working Party 10 on Microbial Corrosion

16. **Guidelines on materials requirements for carbon and low alloy steels for H_2S-containing environments in oil and gas production**
Prepared by Working Party 13 on Corrosion in Oil and Gas Production

17. **Corrosion resistant alloys for oil and gas production: guidance on general requirements and test methods for H2S service**
Prepared by Working Party 13 on Corrosion in Oil and Gas Production

18. **Stainless steel in concrete: state of the art report**
Prepared by Working Party 11 on Corrosion of Steel in Concrete

19. **Sea water corrosion of stainless steels: mechanisms and experiences**
Prepared by Working Party 9 on Marine Corrosion and Working Party 10 on Microbial Corrosion

20. **Organic and inorganic coatings for corrosion prevention: research and experiences**
Papers from EUROCORR '96

21. **Corrosion–deformation interactions**
CDI '96 in conjunction with EUROCORR '96

22. **Aspects of microbially induced corrosion**
Papers from EUROCORR '96 and EFC Working Party 10 on Microbial Corrosion

23. **CO_2 corrosion control in oil and gas production: design considerations**
Prepared by Working Party 13 on Corrosion in Oil and Gas Production
Volumes in the EFC series xxv
Published by Woodhead Publishing Limited, 2013

24. **Electrochemical rehabilitation methods for reinforced concrete structures: a state of the art report**
Prepared by Working Party 11 on Corrosion of Steel in Concrete

25. **Corrosion of reinforcement in concrete: monitoring, prevention and rehabilitation**
Papers from EUROCORR '97

26. **Advances in corrosion control and materials in oil and gas production**
Papers from EUROCORR '97 and EUROCORR '98

27. **Cyclic oxidation of high temperature materials**
Proceedings of an EFC Workshop, Frankfurt/Main, 1999

28. **Electrochemical approach to selected corrosion and corrosion control**
Papers from the 50th ISE Meeting, Pavia, 1999

29. **Microbial corrosion: proceedings of the 4th International EFC Workshop**
Prepared by the Working Party on Microbial Corrosion

30. **Survey of literature on crevice corrosion (1979–1998): mechanisms, test methods and results, practical experience, protective measures and monitoring**
Prepared by F. P. Ijsseling and Working Party 9 on Marine Corrosion

31. **Corrosion of reinforcement in concrete: corrosion mechanisms and corrosion protection**
Papers from EUROCORR '99 and Working Party 11 on Corrosion of Steel in Concrete

32. **Guidelines for the compilation of corrosion cost data and for the calculation of the life cycle cost of corrosion: a working party report**
Prepared by Working Party 13 on Corrosion in Oil and Gas Production

33. **Marine corrosion of stainless steels: testing, selection, experience, protection and monitoring**
Edited by D. Féron on behalf of Working Party 9 on Marine Corrosion

34. **Lifetime modelling of high temperature corrosion processes**
Proceedings of an EFC Workshop 2001
Edited by M. Schütze, W. J. Quadakkers and J. R. Nicholls

35. **Corrosion inhibitors for steel in concrete**
Prepared by B. Elsener with support from a Task Group of Working Party 11 on Corrosion of Steel in Concrete
xxvi Volumes in the EFC series
Published by Woodhead Publishing Limited, 2013

36. **Prediction of long term corrosion behaviour in nuclear waste systems**
Edited by D. Féron on behalf of Working Party 4 on Nuclear Corrosion

37. **Test methods for assessing the susceptibility of prestressing steels to hydrogen induced stress corrosion cracking**
By B. Isecke on behalf of Working Party 11 on Corrosion of Steel in Concrete

38. **Corrosion of reinforcement in concrete: mechanisms, monitoring, inhibitors and rehabilitation techniques**
Edited by M. Raupach, B. Elsener, R. Polder and J.Mietz on behalf of Working Party 11 on Corrosion of Steel in Concrete

39. **The use of corrosion inhibitors in oil and gas production**
Edited by J. W. Palmer, W. Hedges and J. L. Dawson on behalf of Working Party 13 on Corrosion in Oil and Gas Production

40. **Control of corrosion in cooling waters**
Edited by J. D. Harston and F. Ropital on behalf of Working Party 15 on Corrosion in the Refinery Industry

41. **Metal dusting, carburisation and nitridation**
Edited by H. Grabke and M. Schütze on behalf of Working Party 3 on Corrosion by Hot Gases and Combustion Products

42. **Corrosion in refineries**
Edited by J. D. Harston and F. Ropital on behalf of Working Party 15 on Corrosion in the Refinery Industry

43. **The electrochemistry and characteristics of embeddable reference electrodes for concrete**
Prepared by R. Myrdal on behalf of Working Party 11 on Corrosion of Steel in Concrete

44. **The use of electrochemical scanning tunnelling microscopy (EC-STM) in corrosion analysis: reference material and procedural guidelines**
Prepared by R. Lindström, V. Maurice, L. Klein and P. Marcus on behalf of Working Party 6 on Surface Science

45. Local probe techniques for corrosion research
Edited by R. Oltra on behalf of Working Party 8 on Physico-Chemical Methods of Corrosion Testing

46. Amine unit corrosion survey
Edited by J. D. Harston and F. Ropital on behalf of Working Party 15on Corrosion in the Refinery Industry
Volumes in the EFC series xxvii
Published by Woodhead Publishing Limited, 2013

47. Novel approaches to the improvement of high temperature corrosion resistance
Edited by M. Schütze and W. Quadakkers on behalf of Working Party 3 on Corrosion by Hot Gases and Combustion Products

48. Corrosion of metallic heritage artefacts: investigation, conservation and prediction of long term behaviour
Edited by P. Dillmann, G. Béranger, P. Piccardo and H. Matthiesen on behalf of Working Party 4 on Nuclear Corrosion

49. Electrochemistry in light water reactors: reference electrodes, measurement, corrosion and tribocorrosion
Edited by R.-W. Bosch, D. Féron and J.-P. Celis on behalf of Working Party 4 on Nuclear Corrosion

50. Corrosion behaviour and protection of copper and aluminium alloys in seawater
Edited by D. Féron on behalf of Working Party 9 on Marine Corrosion

51. Corrosion issues in light water reactors: stress corrosion cracking
Edited by D. Féron and J-M. Olive on behalf of Working Party 4 on Nuclear Corrosion

52. Progress in corrosion: the first 50 years of the EFC
Edited by P. McIntyre and J. Vogelsang

53. Standardisation of thermal cycling exposure testing
Edited by M. Schütze and M. Malessa on behalf of Working Party 3on Corrosion by Hot Gases and Combustion Products

54. Innovative pre-treatment techniques to prevent corrosion of metallic surfaces
Edited by L. Fedrizzi, H. Terryn and A. Simões on behalf of Working Party 14 on Coatings

55a. Corrosion-under-insulation (CUI) guidelines
Prepared by S. Winnik on behalf of Working Party 13 on Corrosion in Oil and Gas Production and Working Party 15 on Corrosion in the Refinery Industry

55b. Corrosion-under-insulation (CUI) guidelines, Revised 2nd Edition
Prepared by S. Winnik on behalf of Working Party 13 on Corrosion in Oil and Gas Production and Working Party 15 on Corrosion in the Refinery Industry

56. Corrosion monitoring in nuclear systems
Edited by S. Ritter and A. Molander on behalf of Working Party 4 on Nuclear Corrosion

57. **Protective systems for high temperature applications**
Edited by M. Schütze on behalf of Working Party 3 on Corrosion by Hot Gases and Combustion Products
xxviii Volumes in the EFC series
Published by Woodhead Publishing Limited, 2013

58. **Self-healing properties of new surface treatments**
Edited by L. Fedrizzi, W. Fürbeth and F. Montemor on behalf of Working Party 14 on Coatings

59. **Sulphur-assisted corrosion in nuclear disposal systems**
Edited by F. Druyts, D. Féron and B. Kursten on behalf of Working Party 4 on Nuclear Corrosion

60. **Methodology of crevice corrosion testing for stainless steels in natural and treated seawaters**
Edited by U. Kivisäkk, B. Espelid and D. Féron on behalf of Working Party 9 on Marine Corrosion

61. **Inter-laboratory study on electrochemical methods for the characterisation of CoCrMo biomedical alloys in simulated body fluids**
Edited by A. Munoz and S. Mischler on behalf of Working Party 18 on Tribo-Corrosion

62. **Testing tribo-corrosion of passivating materials**
Edited by J-P. Celis and P. Ponthiaux on behalf of Working Party 18 on Tribo-Corrosion

63. **The corrosion performance of metals for the marine environment**
Edited by R. Francis and C. Powell on behalf of Working Party 9 on Marine Corrosion

64. **Recommended practices for corrosion management of pipelines**
Edited by B. Kermani and C. Chevrot on behalf of Working Party 13 on Corrosion in Oil and Gas Production

65. **Corrosion and conservation of cultural heritage metallic artefacts**
Edited by P. Dillmann, D. Watkinson, E. Angelini and A. Adriaens on behalf of Working Party 21 on Corrosion of Archaeological and Historical Artefacts

66. **Understanding Biocorrosing: Fundamentals and Applications**
Edited by T. Liengen, D. Féron, R. Basséguy and I. B. Beech on behalf of Working Party 10 on Microbial Corrosion

67. **Stress Corrosion Cracking of Nickel Based Alloys in Water-Cooled Nuclear Reactors**
Edited by D. Féron and R. Staehle on behalf of Working Party 4 on Nuclear Corrosion

68. **Engineering Tools for Corrosion: Design and Diagnosis**
Prepared by L. Lazzari

Basic Principles

1

1.1 Corrosion reactions

A metal, M, corrodes following a general reaction:

$$M + \text{aggressive environment} = \text{corrosion products} \tag{1.1}$$

If the environment is an electrolyte, the corrosion reaction is the sum of two *electrochemical reactions*:

- an *anodic process* which consists of the oxidation of the metal;
- a *cathodic process* which is a reduction reaction, typically oxygen reduction and hydrogen evolution.

In the case of iron, the anodic reaction is:

$$Fe \rightarrow Fe^{2+} + 2e^- \tag{1.2}$$

where electrons are made available; these electrons are taken by the cathodic reaction, for instance by oxygen:

$$O_2 + 2H_2O + 4e^- \rightarrow 4OH^- \tag{1.3}$$

with production of alkalinity or by hydronium ions in the case of an acidic solution:

$$2H_3O^+ + 2e^- \rightarrow 2H_2O + H_2 \tag{1.4}$$

with consumption of acidity (i.e. production of alkalinity). In addition to above reactions, a corrosion process implies two further processes, namely:

1. an electronic flow within the metal from the anodic area, where electrons are released, to the cathodic region;
2. an ionic current within the electrolyte from the anode to the cathode. Positive ions move in the same direction of the current and negative ions in the opposite one.

Pietro Pedeferri simply summarized a corrosion process as consisting of four processes in series as depicted in Fig. 1.1 [1]. These four processes happen at the same rate because of electroneutrality law: Electrons released by the anodic reaction are timely consumed by the cathodic reaction; therefore:

$$I_a = I_c = I_m = I_{el} = I_{corr} \tag{1.5}$$

This shared current, I_{corr}, measures, in electrochemical units, the corrosion process rate.

Engineering Tools for Corrosion. DOI: http://dx.doi.org/10.1016/B978-0-08-102424-9.00001-X

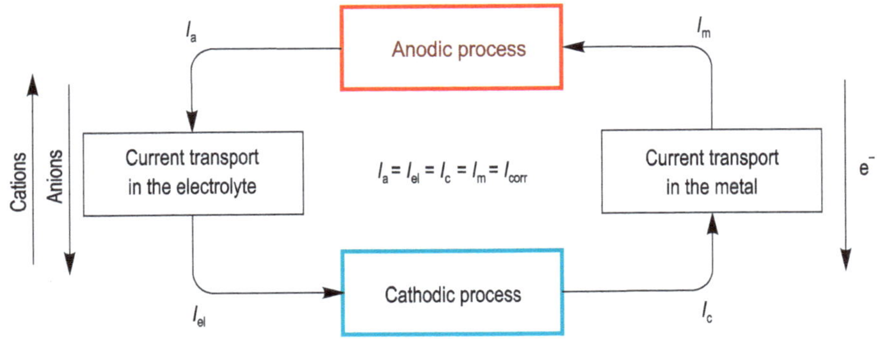

Figure 1.1 Pedeferri's scheme of the electrochemical mechanism of the corrosion process. After Pedeferri P., Corrosione e protezione dei materiali metallici. *Polipress*. Italy: Milan; 2007. ISBN 97888-7398-032-2 (in Italian).

1.2 Electrochemical mechanism

Corrosion process involves various aspects, which deal with anodic and cathodic reactions, electrophoretic migration processes of ionic species and chemical reactions between the metal and products of electrode processes, anodic and cathodic, or between them.

1.2.1 *Anodic processes*

A generic anodic process of a metal is an oxidation reaction, which can produce a metallic ion by releasing electrons:

$$M = M^{z+} + ze \tag{1.6}$$

where z is the number of electrons, M is generic metal and M^{z+} is a metal ion which passes into solution. In specific ranges of pH or in the presence of particular species, insoluble compounds (oxides, hydroxides and salts) can form, for instance an insoluble hydroxide:

$$M + zH_2O = M(OH)_z + zH^+ + ze \tag{1.7}$$

When an insoluble corrosion product forms, so-called *passivation* occurs which reduces the dissolution reaction rate.

1.2.2 *Cathodic processes*

The cathodic process takes the released electrons from the metal to reduce one or more chemical species present in the environment.

The cathodic reactions of practical interest for corrosion are limited in number. In the case of corrosion in an acidic solution, the cathodic process is the reduction of hydronium to produce molecular hydrogen, according to reaction (1.4). In neutral or alkaline environments and in natural environments (soil and waters), the most important cathodic reaction is the oxygen reduction reaction (1.3).

In acidic solutions, the corresponding reaction is the following:

$$O_2 + 4H^+ + 4e = 2H_2O \tag{1.8}$$

The reactant oxygen that appears in above reactions is the molecular oxygen dissolved in the water, the concentration of which varies from 0 to 12 mg/kg of water (mg/kg approximates the concentration in ppm). In addition, water can give a cathodic reaction in specific potential range, as follows:

$$2H_2O + 2e^- \rightarrow 2OH^- + H_2 \tag{1.9}$$

In potable and industrial waters, chlorine reduction is often present:

$$Cl_2 + 2e^- \rightarrow 2Cl^- \tag{1.10}$$

Other cathodic processes are possible; typically, in some accelerated tests, the following is used:

$$Fe^{3+} + e^- \rightarrow Fe^{2+} \tag{1.11}$$

1.3 Stoichiometry (Faraday Law)

As electrons are to be considered chemical species of the electrochemical reactions [2], there is an equivalence between mass and current (Faraday Law):

$$\Delta M = \frac{Q}{F} \cdot \frac{M_W}{z} = \frac{M_W}{zF} \cdot I \cdot t \tag{1.12}$$

where ΔM (g) is the mass loss, Q is the electrical charge (C), F is the Faraday constant (96,485 C or 26.8 A h), z is the valence, I is the current (A), t is the time (s), and M_W is the atomic or molecular mass (g).

The corrosion rate can also be expressed as follows:

$$C_{\text{rate}} = \frac{\Delta M}{\gamma \, A \cdot t} = k \cdot i_a \tag{1.13}$$

where C_{rate} is the corrosion rate in mm/y, i_a is the anodic current density, also called *corrosion current density* in A/m^2, k is a constant of the metal, shown in Table 1.1, which depends on metal density, γ. ΔM is the mass of dissolved metal, A is the exposed surface area, and t is the exposure time.

Table 1.1 Constant k for the equivalence corrosion rate (mm/y) vs corrosion current density (A/m^2)

Metal	Valence	Density (mg/m^3)	Constant k
Iron	Fe^{2+}	7.87	1.17
Nickel	Ni^{2+}	8.90	1.09
Copper	Cu^{2+}	8.96	1.17
Aluminum	Al^{3+}	2.70	1.09
Lead	Pb^{2+}	11.34	2.84
Zinc	Zn^{2+}	7.13	1.50
Tin	Sn^{2+}	7.30	2.67
Titanium	Ti^{2+}	4.51	1.75
Zirconium	Zr^{4+}	6.50	1.91
AISI 304	Fe^{2+}, Cr^{3+}, Ni^{2+}	7.90	1.04
AISI 316	Fe^{2+}, Cr^{3+}, Ni^{2+}, Mo^{3+}	8.00	1.04

Sometimes, corrosion rate is expressed as mass loss adopting the unit mg/m^2 h (mg/dm^2 d is also used in laboratory testing). For iron, the following equivalence applies:

$$1\ \mathrm{mg/m^2\ h} = 1.1\ \mu m/\mathrm{y} \tag{1.14}$$

1.4 Thermodynamic conditions

Thermodynamic condition for a spontaneous corrosion process is $\Delta G < 0$, which becomes the following by introducing the potential:

$$\Delta E > 0; \quad E_C > E_A \tag{1.15}$$

where $\Delta E = E_C - E_A$, E_C is the potential of the cathodic process, and E_A is the potential of the anodic one. Hence, a corrosion process is spontaneous if the potential of the cathodic reaction is more positive (i.e. more noble) than the one of anodic process.

1.4.1 Reference electrodes

The potential is measured by connecting the positive terminal of a voltmeter to the metal (or structure) also said working electrode and the negative terminal to a reference electrode (Fig. 1.2) [3]. By convention, potentials are referred to the standard

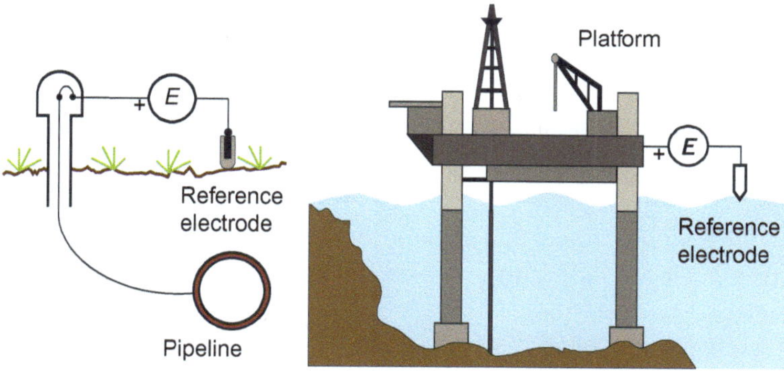

Figure 1.2 Scheme of potential measurement on field.
From L. Lazzari and P. Pedeferri, Cathodic Protection, 2005, Polipress; Italy: Milan, ISBN 88-7398-020-1.

Table 1.2 Reference electrodes

Electrode	Description	Equilibrium reactions	*E* (V SHE)
Standard hydrogen El. (SHE)	H_2(1 atm)|H^+(a = 1)	$2H^+ + 2e^- = H_2$	0
Saturated calomel El. (SCE)	Hg|Hg_2Cl_2, KCl (sat)	$Hg_2Cl_2 + 2e^- = 2Hg + 2Cl^-$	+0.244
Silver/silver chloride (0.1 M)	Ag|AgCl, KCl (0.1 M)	$AgCl + e^- = Ag + Cl^-$	+0.288
Silver/silver chloride/seawater	Ag|AgCl, seawater	$AgCl + e^- = Ag + Cl^-$	+0.250
Copper/copper sulfate (CSE)	Cu|$CuSO_4$ (sat)	$Cu^{2+} + 2e^- = Cu$	+0.318
Zinc/seawater	Zn|seawater	$Zn^{2+} + 2e^- = Zn$	≈ −0.80

hydrogen electrode (SHE) taken as the zero reference. Table 1.2 lists electrodes employed in laboratory and on field.

1.4.2 Potential of anodic reaction, E_a

For a metal dissolution reaction, $M = M^{z+} + ze^-$, equilibrium potential is given by Nernst[1] equation:

$$E_{eq} = E^0 + \frac{RT}{zF} \ln \frac{a_{M^{z+}}}{a_M} \tag{1.16}$$

where E^0 (V SHE) is the standard potential of metal, M; $a_{M^{z+}}$ (mol/L) is the ion concentration; a_M is the metal concentration taken as unitary ($a_M = 1$). Table 1.3 reports standard potentials for most common reactions.

[1] Walther Herman Nernst (1864–1941) was a German chemist. He received Nobel Prize in chemistry for 1920.

For alloys, standard potential, E^0, is calculated as weighted average of standard potential of components; for instance for 18/8 stainless steel (AISI 304 grade), standard potential is $E^0_{SS} = 0.74 \cdot E^0_{Fe} + 0.18 \cdot E^0_{Cr} + 0.08 \cdot E^0_{Ni} = -0.4$ V SHE.

Nernst equation shorts to:

$$E_{eq} = E^0 + \frac{0.059}{z} \log a_{M^{z+}} \tag{1.17}$$

where 0.059 V is the term 2.3 *RT*/F at 25°C (2.3 is conversion coefficient from natural to decimal-base logarithm).

Metal ion concentration $a_{M^{z+}}$ is assumed 10^{-6} mol/L, as suggested by Pourbaix[2] [4], for electrolytes not containing metal ions, as in the case of metals exposed to waters or immersed in soil. The equilibrium potential shortens to:

$$E_{eq} = E^0 - \frac{0.354}{z} \tag{1.18}$$

When insoluble products and complexing species form, for example oxides, hydroxides, or sulfides, metal ion concentration drops to extremely low value, hence allowing the equilibrium potential to match thermodynamic condition for corrosion. An example is copper dissolution in deaerated, concentrated hydrochloric acid, where copper forms a complex $Cu^{2+} + 4Cl^- = Cu(Cl)_4^{2-}$ with constant of the order 10^{20}, which gives a copper ion concentration of the order of $(Cu^{2+}) \cong 10^{-20}$; equilibrium potential of copper drops to -0.25 V SHE allowing acidic corrosion.

1.4.3 Potential of cathodic reaction, E_c

Cathodic processes rank from the most noble downward as the following list:

- Chlorine reduction to give chloride ($E^0 = +1.36$);
- Oxygen reduction ($E = +1.23 - 0.059$ pH);
- Ferric to ferrous ion reduction ($E^0 = +0.77$);
- Copper ions reduction ($E^0 = +0.34$);
- Hydronium reduction ($E = -0.059$ pH);

where potential is expressed in V vs SHE.

1.4.3.1 Hydrogen evolution

The hydrogen evolution process in acidic solutions is given by reactions (1.4) and (1.9) which are energetically equivalent and therefore characterized by the same

[2] Marcel Pourbaix (1904–1998) was a Belgium electrochemist who proposed and developed *E*-pH diagrams which now have his name.

Table 1.3 Electrochemical series of standard potentials

Electrode reactions	*E* V vs SHE	Electrode reactions	*E* V vs SHE
$F_2 + 2H^+ + e^- = 2HF$	+3.03	$Cu^{2+} + e^- = Cu^+$	+0.158
$O_3 + 2H^+ + e^- = O_2 + H_2O$	+2.07	$Sn^{4+} + 2e^- = Sn^{2+}$	+0.15
$Co^{3+} + 3e^- = Co$	+1.842	$2H^+ + 2e^- = H_2$	0
$Au^+ + e^- = Au$	+1.68	$2D^+ + 2e^- = D_2$	−0.0034
$Mn^{3+} + e^- = Mn^{2+}$	+1.51	$Fe^{3+} + 3e^- = Fe$	−0.036
$Au^{3+} + 3e^- = Au$	+1.50	$Pb^{2+} + 2e^- = Pb$	−0.1263
$MnO_4^- + 8H^+ + 5e^- = Mn^{2+} + 4H_2O$	+1.491	$Sn^{2+} + 2e^- = Sn$	−0.1364
$PbO_2 + 4H^+ + 2e^- = Pb^{2+} + 2H_2O$	+1.467	$Ge^{4+} + 4e^- = Ge$	−0.15
$Cl_2 + 2e^- = 2Cl^-$	+1.3583	$Mo^{3+} + 3e^- = Mo$	−0.20
$Cr_2O_7^{2-} + 14H^+ + 6e^- = 2Cr^{3+} + 7H_2O$	+1.33	$Ni^{2+} + 2e^- = Ni$	−0.25
$O_2 + 4H^+ + 4e^- = 2H_2O$	+1.23	$Co^{2+} + 2e^- = Co$	−0.28
$CrO_4^{2-} + 8H^+ + 3e^- = Cr^{3+} + 4H_2O$	+1.195	$Mn^{3+} + 3e^- = Mn$	−0.283
$Pt^{2+} + 3e^- = Pt$	+1.19	$In^{3+} + 3e^- = In$	−0.342
$Br_2 + 2e^- = 2Br^-$	+1.087	$Cd^{2+} + 2e^- = Cd$	−0.40
$HNO_3 + 3H^+ + 3e^- = NO + 2H_2O$	+0.96	$Cr^{3+} + e^- = Cr^{2+}$	−0.41
$2Hg^{2+} + 2e^- = Hg_2^{2+}$	+0.92	$Fe^{2+} + 2e^- = Fe$	−0.44
$Hg^{2+} + 2e^- = 2Hg$	+0.851	$Cr^{3+} + 3e^- = Cr$	−0.74
$Ag^+ + e^- = Ag$	+0.7996	$Zn^{2+} + 2e^- = Zn$	−0.76
$Hg_2^{2+} + 2e^- = 2Hg$	+0.7961	$V^{3+} + 3e^- = V$	−0.876
$Fe^{3+} + e^- = Fe^{2+}$	+0.770	$Cr^{2+} + 2e^- = Cr$	−0.913
$O_2 + 2H^+ + 2e^- = H_2O$	+0.682	$Nb^{3+} + 3e^- = Nb$	−1.10
$Hg_2SO_4 + 2e^- = 2Hg + SO_4^{2-}$	+0.62	$Mn^{2+} + 2e^- = Mn$	−1.18
$MnO_4^- + 2H_2O + 3e^- = MnO_2 + 4\ OH^-$	+0.588	$V^{2+} + 2e^- = V$	−1.18
$I_2 + 2e^- = 2I^-$	+0.534	$Ti^{3+} + 3e^- = Ti$	−1.21
$Cu^+ + e^- = Cu$	+0.522	$Zr^{4+} + 4e^- = Zr$	−1.53
$2NO_2^- + 4H_2O + 6e^- = N_2 + 8OH^-$	+0.42	$Ti^{2+} + 2e^- = Ti$	−1.63
$SO_4^{2-} + 6e^- + 8H^+ = S + 4H_2O$	+0.36	$Al^{3+} + 3e^- = Al$	−1.66
$Cu^{2+} + 2e^- = Cu$	+0.34	$Mg^{2+} + 2e^- = Mg$	−2.36
$2NO_3^- + 6H_2O + 10e^- = N_2 + 12OH^-$	+0.25	$Na^+ + e^- = Na$	−2.71
$AgCl + e^- = Ag + Cl^-$	+0.22	$Ca^{2+} + 2e^- = Ca$	−2.86
$SO_4^{2-} + 2e^- + 2H^+ = SO_3^{2-} + H_2O$	+0.17	$Li^+ + e^- = Li$	−3.05

equilibrium potential. Assuming $pH_2 = 1$ bar and $a_{H_2O} = 1$, according to the Nernst Law, the equilibrium potential varies with pH as follows:

$$E_{eq,H_2} = 0 + 0.059 \log a_{H^+} = -0.83 - 0.059 \log a_{OH^-} = -0.059\text{pH} \quad (1.19)$$

In the Pourbaix diagram, E-pH, equilibrium potential of hydrogen evolution is a straight line having a slope of -0.059 V/decade.

1.4.3.2 Oxygen reduction

In neutral or alkaline aerated solutions as well as in aerated acidic solutions, the oxygen reduction reaction are (1.3) and (1.8) characterized by the same equilibrium potential. With reference to the standard conditions, $pO_2 = 1$ bar, $a_{H_2O} = 1$ and 25°C, the equilibrium potential is given by the Nernst's equation:

$$E_{eq,O_2} = 0.40 - 0.059 \log a_{OH^-} = 1.23 + 0.059 \log a_{H^+} = 1.23 - 0.059 \text{ pH} \quad (1.20)$$

In the Pourbaix diagram, E-pH, equilibrium potential of oxygen reduction is a straight line which is parallel to that of hydrogen evolution with same slope, 0.059 V/decade. Equilibrium potential changes with oxygen content, which determines the effective oxygen partial pressure for the input in Nernst equation; accordingly, introducing the Henry Law for oxygen solubility, the following empirical equation can be used:

$$E_{O_2} \cong 1.23 - 0.33 \log\left(\frac{50}{[O_2]}\right) - 0.059 \text{ pH} \quad (1.21)$$

where $[O_2]$ is the oxygen content in ppm. From Eq. (1.21), it results that a variation of 1 ppm of the oxygen content determines an average change of the equilibrium potential of about 50 mV for low oxygen content (i.e. below 5 ppm).

1.5 Kinetics of aqueous corrosion

In principle, corrosion rate is given by the formal equation: *driving voltage/reaction resistance*, where the numerator is clearly known (i.e. $E_C - E_A$), whereas the denominator is almost variable. To represent this relationship, a potential-current diagram can be adopted as shown in Fig. 1.3, which depicts the most used one in semilogarithm scale, E-log$i$, universally known as *Evans diagram*[3]. The reaction resistance is the slope of the curve and is the sum of three contributions, namely:

- anodic overvoltage, i.e. energy dissipation associated with metal dissolution or oxidation;
- cathodic overvoltage, i.e. energy dissipation associated with cathodic process;
- ohmic drop in the electrolyte (the one in the metal is negligible).

[3] Ulick Richardson Evans (1889–1980) was a British scientist who was described in the Biographical Memoirs of Fellows of the Royal Society as the "Father of the modern science of corrosion and protection of metals," contributing on a firm foundation of electrochemical nature of corrosion. His book "Corrosion of Metals", published in 1924, remains a milestone in the history of corrosion science.

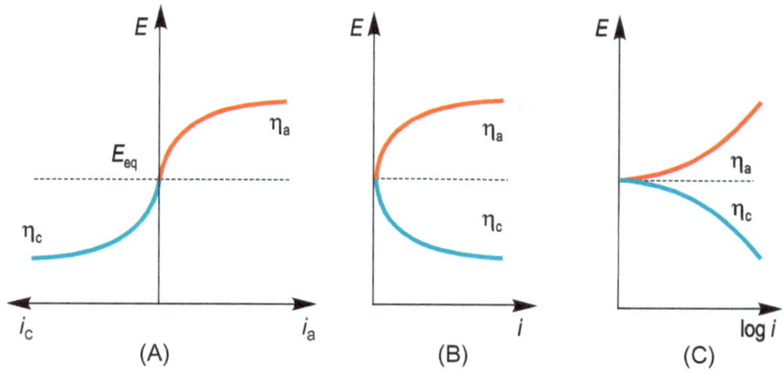

Figure 1.3 Representation of overvoltage.
From Pedeferri P., Corrosione e protezione dei materiali metallici. Polipress. Italy: Milan; 2007. ISBN 97888-7398-032-2 (in Italian).

1.5.1 Anodic overvoltage

Piontelli[4] showed that anodic overvoltage associated with metal dissolution depends on the nature of the metal and proposed the following classification:

- *normal* or low overvoltage metals, for which even at high current density, from both the anode side and the cathode side, the overvoltage, η_M, is less than 10 mV. This class includes the *low*-melting temperature metals: Cd, Hg, Sn, Pb, Mg, Al and Zn;
- *inert* or high overvoltage metals, for which even at small current density and in a wide range of conditions, in the absence of polarizing current, overvoltage, η_M, is greater than 100 mV. This class includes the *high*-melting temperature metals: Fe, Co, Ni, Cr, Mo, Ti and metals of the platinum and transition groups;
- *intermediate* metals, for which η_M is between the two above-mentioned limits. Examples of intermediate metals are: Cu, Au, and Ag.

For intermediate and inert metals, the overvoltage-to-current density relationship follows *Tafel Law*[5]:

$$\eta_a = b_a \log \frac{i}{i_0} \tag{1.22}$$

where overvoltage is in V; i is the current density (mA/m^2); i_0 is the exchange current density (mA/m^2); b_a is the module of Tafel slope (V/decade). Relevant values are reported in Table 1.4.

[4] Roberto Piontelli (1909–1971) was an eminent Italian electrochemist.

[5] Julius Tafel (1862–1918) was a German chemist who worked in electrochemistry with Wilhelm Ostwald.

Table 1.4 Tafel slope, *b*, and order of magnitude of exchange current density, i_0, for metals

Metals	Tafel slope, *b* (mV/decade)	Exchange current density, i_0 (order of magnitude in mA/m²)
Inert	Monovalent 120 Bivalent 60	1
Intermediate	Monovalent 120 Bivalent 60	10
Normal	$b \cong 0$	10^4

From Pedeferri P., Corrosione e protezione dei materiali metallici. Polipress. Italy: Milan; 2007. ISBN 97888-7398-032-2 (in Italian).

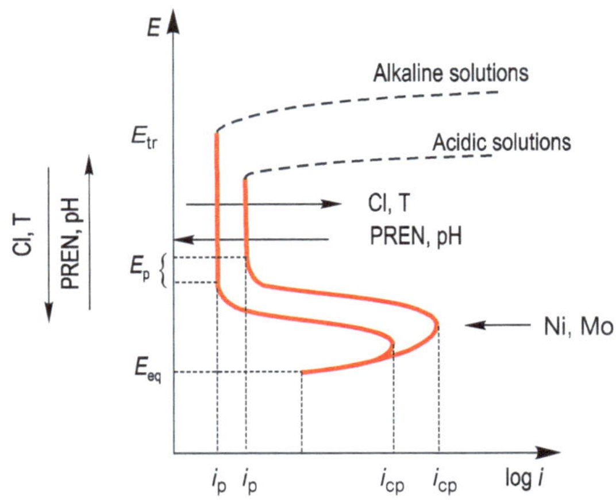

Figure 1.4 Anodic overvoltage for stainless steels showing influencing parameters. From Pedeferri P., Corrosione e protezione dei materiali metallici. Polipress. Italy: Milan; 2007. ISBN 97888-7398-032-2 (in Italian).

1.5.1.1 Passivity-related parameters for stainless steels

When a metal passivates, like Al, Ti, Fe, and stainless steels, anodic overvoltage is so high that metal dissolution rate practically zeros. The shape of the anodic curve changes as shown in Fig. 1.4 and is characterized by a number of parameters as follows:

- *critical passivation current density*, i_{cp};
- *passivity current density*, i_p;
- *passivity potential* (or Flade potential), E_p;
- *transpassivity potential*, E_{tr};
- *pitting potential*, E_{pit};
- *passivity interval*, $(E_{tr} - E_p)$.

A variety of factors influences parameters and shape of the anodic curve: primarily the composition (pure metal or alloy) but also surface finishing and metallurgical condition, i.e., annealed or cold worked. Moreover, environmental properties, such as temperature, acidity, chlorides concentration, influence the curve shape as shown in Fig. 1.4.

E_p is the *passivity potential*, also called *Flade Potential*[6]. From experimental data, passivity potential, E_p, for stainless steels can be expressed as function of pH, PREN, chloride concentration, and temperature:

$$E_p \cong \left(\frac{298}{T+273}\right)\cdot\left[0.63+0.06\left(\frac{pH-7}{5}\right)-\frac{PREN}{100}-0.05\log\left(\frac{Cl}{36}+1\right)\right] \quad (1.23)$$

where E_p is in V SHE; 0.63 (V) is the Flade constant; Cl (ppm) is the chloride content; PREN is a stainless steel related composition parameter calculated by the following expression: PREN = %Cr + 3.3%Mo + 16%N for austenitic stainless steels and PREN = %Cr + 3.3%Mo + 30%N for duplex stainless steels; T (°C) is the temperature.

Critical passivation current density, i_{cp}, for stainless steels can be expressed as function of pH, temperature and the sum of Ni and Mo contents as follows:

$$i_{cp} \cong 10^{-\left(0.1(\%Ni+\%Mo)+\frac{pH-7}{5}\right)}\cdot 2^{\frac{T-25}{50}} \quad (1.24)$$

where i_{cp} is in A/m^2.

Passivity current density, i_p (mA/m^2), can be expressed as parametric empirical function of PREN, pH, chloride content and temperature, as follows:

$$i_p \cong k\cdot 10^{\left[0.5\log\left(\frac{Cl}{36}+1\right)-\frac{PREN}{13}-\frac{pH-7}{5}\right]}\cdot 2^{\frac{T-25}{50}} \quad (1.25)$$

where i_p is in mA/m^2; (Cl^-) is the chloride concentration (ppm), k (mA/m^2) is an experimental constant (close to 1).

1.5.2 Cathodic overvoltage

Cathodic overvoltage depends on cathodic reaction. *Activation overvoltage* of both hydrogen evolution and oxygen reduction reactions follows Tafel Law as below:

$$\eta_c = b_c \log\frac{i}{i_0} \quad (1.26)$$

[6] This parameter was defined by Friedrich Flade (1880–1916) a German electrochemist, beginning 20th century and represents the potential plateau recorded on potential decay test on iron passivated by anodizing in 1N H_2SO_4 (pH = 0). Flade potential depends on pH as follows: E_F (V SHE) $\cong 0.63 - n\cdot 0.059$pH, where n is a constant, which assumes a value between 1 and 2.

where i_0 is *exchange current density*. Tafel Law can be derived from *Butler–Volmer equation*[7]:

$$i = i_0\left(e^{\frac{(1-\beta)zF\eta}{RT}} - e^{\frac{-\beta zF\eta}{RT}}\right) \tag{1.27}$$

where η is the overvoltage given by $\eta = E - E_{eq}$, i_0 is the exchange current density, β is the transfer coefficient often taken as 0.5, F is the Faraday constant, and z is the reaction equivalence. The Butler–Volmer equation states that the exchange of current on the surface of an electrode takes place only if an activation energy is exceeded.

1.5.2.1 Hydrogen evolution

Hydrogen overvoltage, η_H, follows Tafel Law with a slope of 0.12 V/decade, therefore:

$$\eta_H = 0.12 \log \frac{i}{i_{0,H}} \tag{1.28}$$

where overvoltage is in V, and $i_{0,H}$ is the exchange current density of hydrogen evolution, which depends on metal as Table 1.5 reports. It appears the following:

- *low-melting* temperature metals have low dissolution-related overvoltage and high hydrogen evolution overvoltage;
- *high-melting* temperature metals have high dissolution-related overvoltage and low hydrogen evolution overvoltage.

For this reason, Piontelli suggested to name this relationship as "*reverse-correlation*," as shown in Fig. 1.5. The strong influence of metal nature on hydrogen overvoltage explains why metallic impurities influence corrosion rate in acidic solutions; for instance pure Zn, Pb and Al (i.e. low melting temperature metals with high hydrogen overvoltage), show a high corrosion rate in the presence of high melting temperature impurities, like Fe and Ni, because having low hydrogen overvoltage.

From Eq. (1.28), corrosion rate can be derived when anodic current density (i.e. corrosion current density, i_{corr}) equals the cathodic current density of the hydrogen evolution. In short:

$$i_{corr} = i_H = i_{0,H} \cdot 10^{\frac{\eta_H}{0.12}} \tag{1.29}$$

where symbols are known.

[7] John Alfred Valentine Butler (1899–1977) was an English physical chemist who developed kinetic theories of the origin of electrode potentials and developed the general theory of overvoltage with hydrogen and oxygen electrodes. Max Volmer (1885–1965) was a German physical chemist, who made important contributions in electrochemistry, in particular on electrode kinetics. He codeveloped the Butler–Volmer equation.

Table 1.5 Exchange current density, $i_{o,H}$, on metal, M, and exchange current density, $i_{o,M}$, of metal M

Metal classification		Exchange current density, $i_{o,H}$, on metal, M (order of magnitude in mA/m^2)	Exchange current density, $i_{o,M}$ (order of magnitude in mA/m^2)
Inert	Pt	10^4	1
	Pd, Rh	10^3	
	W, Ta	10^2	
	Co, Ni	10	
	Fe, Cu	1	
Intermediate	Ag, Be	10^{-1}	10
Normal	Sn, Al	10^{-2}	$\approx 10^4$
	Zn	10^{-3}	
	Pb, Hg	10^{-4}	

From Pedeferri P., Corrosione e protezione dei materiali metallici. Polipress. Italy: Milan; 2007. ISBN 97888-7398-032-2 (in Italian).

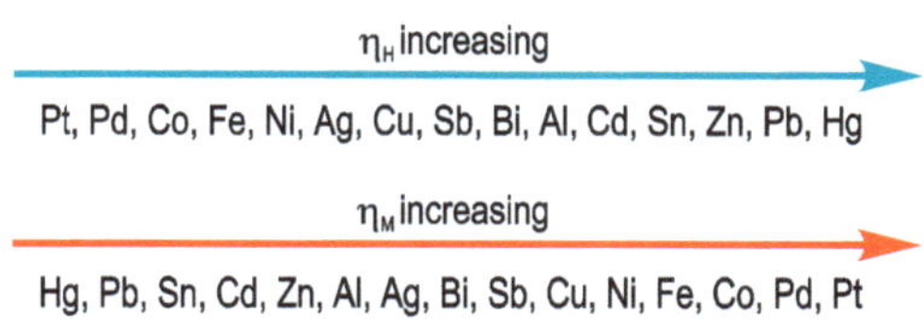

Figure 1.5 Correlation between metal dissolution overvoltage and hydrogen evolution overvoltage, provided by Piontelli [2].
From Pedeferri P., Corrosione e protezione dei materiali metallici. Polipress. Italy: Milan; 2007. ISBN 97888-7398-032-2 (in Italian).

1.5.2.2 Oxygen reduction

Oxygen reduction process involves two dissipation contributions, one corresponding to the process of charge transfer to the metal surface (activation overvoltage), and the other one to the transport of oxygen in the solution (diffusion overvoltage).

Activation overvoltage depends intrinsically on metal and pH and follows Tafel Law as shown in Fig. 1.6. It is low for gold, palladium, platinum, and graphite and much higher for chromium, stainless steels, titanium, and zirconium where an oxide film forms.

Diffusion overvoltage sets up because of changes in chemical composition in the electrolyte as consequence of oxygen diffusion. Fig. 1.7 illustrates how oxygen concentration varies. The concentration gradient of the oxygen within the diffusion layer gives rise to an overvoltage contribution, also called *concentration polarization*, η_{conc,O_2}, which is given in absolute value by:

$$\eta_{conc,O_2} = \frac{RT}{zF} \ln \frac{C_1}{C_2} \tag{1.30}$$

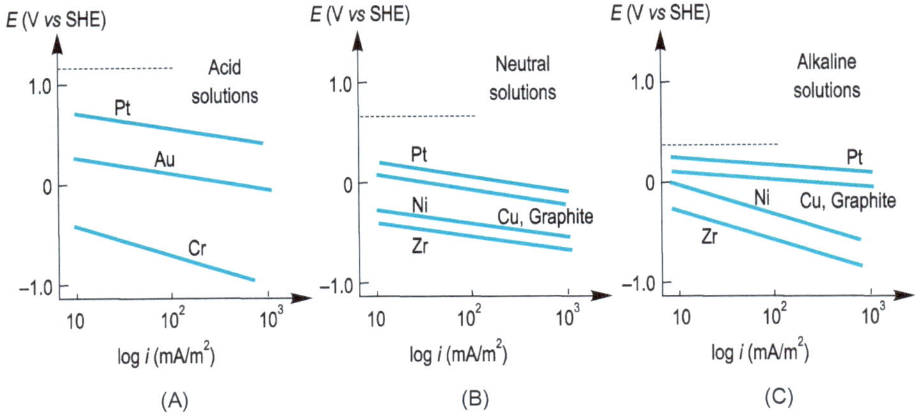

Figure 1.6 Charge transfer overvoltage of oxygen reduction: (A) acidic, (B) neutral and (C) alkaline solutions.
From Pedeferri P., Corrosione e protezione dei materiali metallici. Polipress. Italy: Milan; 2007. ISBN 97888-7398-032-2 (in Italian).

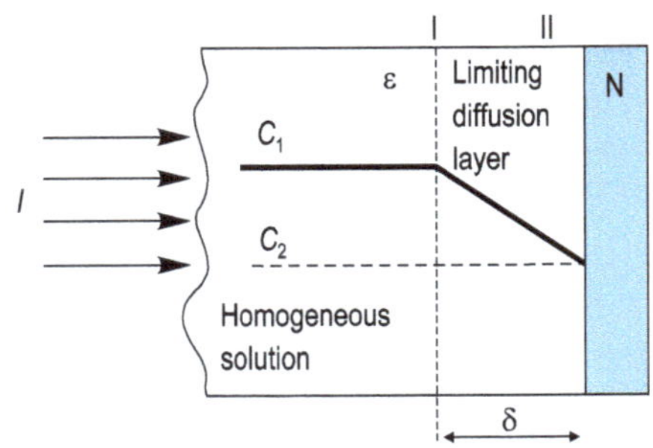

Figure 1.7 Diffusion process representation.
From Pedeferri P., Corrosione e protezione dei materiali metallici. Polipress. Italy: Milan; 2007. ISBN 97888-7398-032-2 (in Italian).

where C_1 is the oxygen concentration in bulk, and C_2 is the oxygen concentration at metal surface. As C_2 zeros, η_{conc,O_2} increases up to a maximum of about 1 V, because hydrogen evolution begins when potential drops below the hydrogen evolution potential. In stationary conditions, oxygen diffusion rate reaches a maximum, called *oxygen limiting current density*, i_L, as stated by Fick Law:

$$i_L = 4\,F\frac{D}{\delta}C_1 \tag{1.31}$$

where i_L is in mA/m^2; D (m^2/s) is the diffusion coefficient, C_1 (mol/L) is the oxygen concentration in bulk, δ (m) is the diffusion layer thickness, F is the Faraday

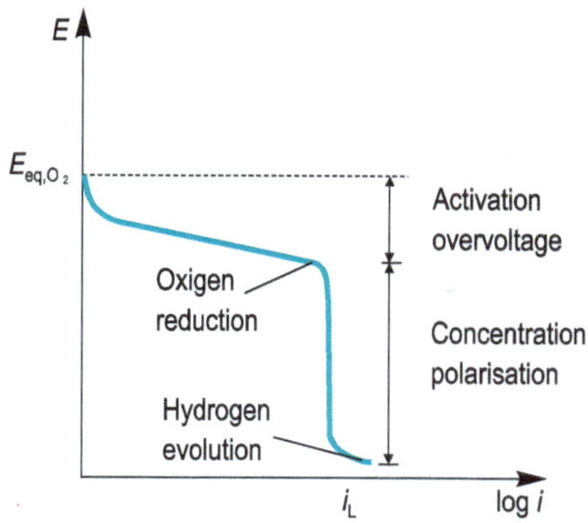

Figure 1.8 Overvoltage contributes for oxygen reduction and hydrogen evolution.
From Pedeferri P., Corrosione e protezione dei materiali metallici. Polipress. Italy: Milan; 2007. ISBN 97888-7398-032-2 (in Italian).

constant (96,485 C). The absolute value of overvoltage concentration, η_{conc,O_2}, becomes:

$$\eta_{conc,O_2} = \frac{RT}{4F} \ln \frac{C_2}{C_1} = \frac{RT}{4F} \ln \frac{i_L - i}{i_L} = 0.015 \log \frac{i_L - i}{i_L} \tag{1.32}$$

Overall, oxygen reduction overvoltage is the sum of activation overvoltage and concentration polarization, as follows (absolute value):

$$\eta_{O_2} = \eta_{att,O_2} + \eta_{conc,O_2} = b \log \frac{i}{i_{0,O_2}} + 0.015 \log \frac{i_L - i}{i_L} \tag{1.33}$$

$$\eta_{O_2} = 1.23 + 0.12 \log \frac{i - i_L}{i_{0,H}} \tag{1.34}$$

where i_{0,O_2} (mA/m^2) is the exchange current density for oxygen reduction, which depends on metal. Tafel slope of oxygen reduction can be taken as the one for hydrogen evolution as 0.12 V/decade, $i_{0,H}$ (mA/m^2) is the exchange current density for hydrogen evolution, and i_L is the oxygen limiting current density.

In summary, as shown in Fig. 1.8, cathodic curve is characterized by three intervals: $i < i_L$ where activation overvoltage prevails; i approaches i_L ($i \cong i_L$) under diffusion control and third interval, $i > i_L$, when hydrogen evolution becomes predominant.

1.5.2.3 Oxygen limiting current density

When anodic current density (i.e., corrosion current density, i_{corr}) equals the cathodic current density of oxygen reduction in condition under diffusion control, corrosion rate coincides with oxygen limiting current density governed by the Fick Law as per Eq. (1.30). Influencing parameters varies as follows:

- diffusion layer thickness, δ, in stagnant condition in the range 1−3 mm (oxygen content from 1 to 11 ppm) and a parabolic variation with water velocity, v;
- the diffusion coefficient D (m^2/s) is $2.25 \cdot 10^{-9}$ at 25°C (1.97 at 20°C; 4.82 at 60°C) [Log D (cm^2/s) = $-4.410 + 773.8/T - (506.4/T)^2$] varies approximately with temperature by doubling every 25°C increase.

Accordingly, the oxygen limiting current density, i_L can be expressed by the following rearranged equation:

$$i_L \cong 10 \cdot 2^{\frac{T-25}{25}} \cdot [O_2] \cdot \left(1 + \sqrt{v}\right) \tag{1.35}$$

where i_L is in mA/m^2; $[O_2]$ (mg/L ≈ ppm) is the oxygen content in water; v (m/s) is the water velocity, and T (°C) is the temperature.

1.5.2.4 Overall cathodic current

In acidic, aerated solutions both cathodic reactions, i.e. oxygen reduction and hydrogen evolution, can take place, depending on metal involved. An overall equation can be derived from Eq. (1.34), leading to the following:

$$i = i_L + i_{0,H}\ e^{\frac{-\eta_H}{b_H}} \tag{1.36}$$

where i (mA/m^2) is the total cathodic current density; i_L (mA/m^2) is the oxygen limiting current density; $i_{0,H}$ (A/m^2) is the exchange current density for hydrogen evolution; η_H (V) is the overvoltage given by $E - E_{eq,H}$; b_H (V/decade) is the Tafel slope of hydrogen evolution reaction.

1.6 Summary

Table 1.6 summarizes equations proposed for the calculation of basic thermodynamic and kinetic parameters of corrosion processes.

Table 1.6 Equations for the calculation of basic thermodynamic and kinetic parameters of corrosion processes

	Equations	Parameters
Stoichiometry; Faraday Law	$\Delta M = \frac{Q}{F} \cdot \frac{M_W}{z} = \frac{M_W}{zF} \cdot I \cdot t$ $C_{rate} = \frac{\Delta M}{\gamma A \cdot t} = k\, i_a$ For iron: 1 mg/m^2 h = 1.1 μm/y 1 mA/m^2 = 1.17 μm/y	C_{rate} (mm/y) is corrosion rate in i_a (A/m^2) anodic current density ΔM (kg) mass of dissolved metal A (m^2) exposed surface area t (y) exposure time γ (mg/m^3) metal density k metal constant (Table 2.1) E^0 (V SHE) standard potential a_M^{z+} (mol/L) metal ion conc. z valence of metal ions F (C) Faraday Constant $[O_2]$ (ppm) oxygen concentration
Equilibrium potential for metal dissolution (V SHE)	$E_{eq} = E^0 + \frac{0.059}{z} \log a_{M^{z+}}$	
Equilibrium potential for cathodic processes (V SHE)	$E_{eq,H_2} = -0.059\ pH$ $E_{eq,O_2} = 1.23 - 0.059\ pH$ $E_{O_2} \cong 1.23 - 0.33 \log\left(\frac{50}{[O_2]}\right) - 0.59 pH$	
Butler–Volmer equation	$i = i_0\left(e^{\frac{(1-\beta)\, zF\eta}{RT}} - e^{\frac{-\beta\, zF\, \eta}{RT}}\right)$	β transfer coefficient D (m^2/s) diffusion coefficient δ (m) diffusion layer thickness F (C) Faraday Constant η (V) overvoltage η_a (V) anodic overvoltage η_c (V) cathodic overvoltage η_{conc,O_2} (V) oxygen overvoltage η_H (V) hydrogen overvoltage
Tafel Law	$\eta_a = b_a \log \frac{i}{i_0};\quad \eta_c = b_c \log \frac{i}{i_0}$	
Hydrogen overvoltage (V)	$\eta_H = 0.12 \log \frac{i}{i_{0,H}}$	
Hydrogen evolution current density (A/m^2)	$i_{corr} = i_H = i_{0,H} \cdot 10^{\frac{\eta_H}{0.12}}$	

(*Continued*)

Table 1.6 (Continued)

	Equations	Parameters
Oxygen overall overvoltage (V)	$\eta_{O_2} = 1.23 + 0.12 \log \frac{i - i_L}{i_{0,H}}$	i (A/m^2) current density i_0 (A/m^2) exchange current density i_{corr} (A/m^2) corrosion rate i_L (mA/m^2) oxygen limiting c.d. $[O_2]$ (ppm) oxygen content R (J/mol K) gas constant T (K) temperature T (°C) temperature v (m/s) velocity z valence of metal ions
Oxygen limiting current density	$i_L = 4\,F \frac{D}{\delta} C_1$ $i_L \cong 10 \cdot 2^{\frac{T-25}{25}} \cdot [O_2] \cdot (1 + \sqrt{v})$	
Overall cathodic current	$i = i_L + i_{0,H}\, e^{-\frac{\eta_H}{b_H}}$	
Passivity potential of stainless steels (V SHE)	$E_p \cong \left(\frac{298}{T+273}\right) \cdot \left[0.63 + 0.06\left(\frac{pH-7}{5}\right) - \frac{PREN}{100} - 0.05 \log\left(\frac{Cl}{36} + 1\right)\right]$	
Critical passivation current density (A/m^2) for stainless steels	$i_{cp} \cong 10^{-\left(0.1(\%Ni+\%Mo) + \frac{pH-7}{5}\right)} \cdot 2^{\frac{T-25}{50}}$	PREN (Cr% + 3.3Mo% + 16 − 30%N) T (°C) temperature [Cl] (ppm) chloride content k experimental constant (≈1 mA/m^2)
Passivity current density (mA/m^2) for stainless steels	$i_p \cong k\; 10^{\left[0.5 \cdot \log\left(\frac{Cl}{36}+1\right) - \frac{PREN}{13} - \frac{pH-7}{5}\right]} \cdot 2^{\frac{T-25}{50}}$	

1.7 Appendix

Table 1.A1 reports the passivity related parameters, E_p, i_{cp}, and i_p, as obtained with proposed equations, for most used stainless steels in industry, operating at different temperatures and pH.

1.7.1 Case study—design parameters for an anodic protection system

Anodic protection is an electrochemical technique for maintaining passive a metal with an active–passive behavior when placed in an aggressive environment. With reference to Fig. 1.A1, by applying an anodic polarization, potential is moved from point ① of free corrosion condition to point ② within the passive interval and maintained in it, provided that initial anodic current density exceeds the critical passivation current density, i_{cp}, then protection current drops to the passivity current density, i_p.

Anodic protection feasibility depends on the anodic curve characteristics of the metal, especially on the following key parameters:

- critical passivation current density, i_{cp}, which is the current that the system should be able to supply in order to bring the material to passive conditions;
- passive current density, i_p, which measures the corrosion rate once passivity is established;
- passivation interval that defines the operating potential range ($E_p < E < E_p + 300$ mV).

Obviously, both passive current density, i_p, and critical passivation current density, i_{cp}, depend on metal and environment as far as type, composition, and temperature are concerned. For example, in most environments, titanium passivates better than steel because of its low values of i_{cp} and i_p. Austenitic stainless steels, and particularly those containing molybdenum, passivate better than ferritic stainless steels because, like titanium, their i_{cp} is lower [3].

The most interesting field of application is the storage of concentrated sulfuric acid, where anodic protection reduces the corrosion rate by two orders of magnitude, thereby reducing metal contamination and hydrogen evolution.

The design of the anodic protection of a storage tank made of austenitic stainless steel for sulfuric acid, requires the knowledge of above key-parameters to determine the size of the potentiostat. For AISI 316 stainless steel tank, with total surface area of 100 m^2, containing concentrated sulfuric acid (67%) key parameters are as follows:

- E_p, given by Eq. (1.23): pH = 0; T = 25°C; E_p = 0.27 V SHE (0.03 V SCE);
- i_{cp}, given by Eq. (1.24): pH = 0; T = 25°C; 2.5 A/m^2;
- i_p, given by Eq. (1.25): pH = 0; T = 25°C; 0.6 mA/m^2.

It results that initial current would be 2500 A, operating current 0.6 A and protection interval would be 0.2–0.4 V SCE.

Table 1.A1 Calculated passivity-related parameters for most used stainless steels (Cl content below 100 ppm) at room temperature, 25°C and 80°C

PREN	pH = 1; T = 25°C			pH = 1; T = 80°C			pH = 7; T = 25°C			pH = 13; T = 80°C		
	E_p	i_{cp}	i_p	E_p	i_{cp}	i_p	E_p	i_{cp}	i_p	E_p	i_{cp}	i_p
	V SHE	(A/m^2)	(mA/m^2)	V SHE	(A/m^2)	(mA/m^2)	V SHE	(A/m^2)	(mA/m^2)	V SHE	(A/m^2)	(μA/m^2)
13	0.40	16	3	0.34	34	6.6	0.47	1.0	0.19	0.46	0.14	26
18	0.36	2.5	1.3	0.29	0.96	2.7	0.42	0.2	0.08	0.42	0.02	11
25	0.28	1.6	0.4	0.24	0.81	0.8	0.35	0.1	0.02	0.36	0.01	3
35	0.18	0.2	0.06	0.15	0.69	0.2	0.25	<0.1	<0.01	0.27	<0.01	1
40	0.13	<0.1	0.02	0.11	0.58	<0.1	0.20	<0.1	<0.01	0.23	<0.01	<1

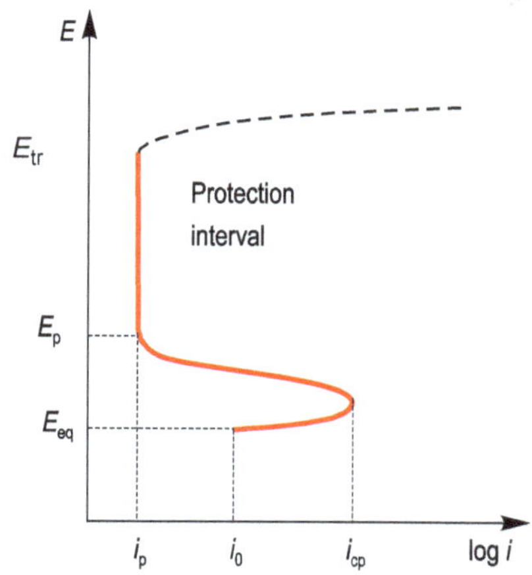

Figure 1.A1 Parameters and operating conditions in anodic protection systems.

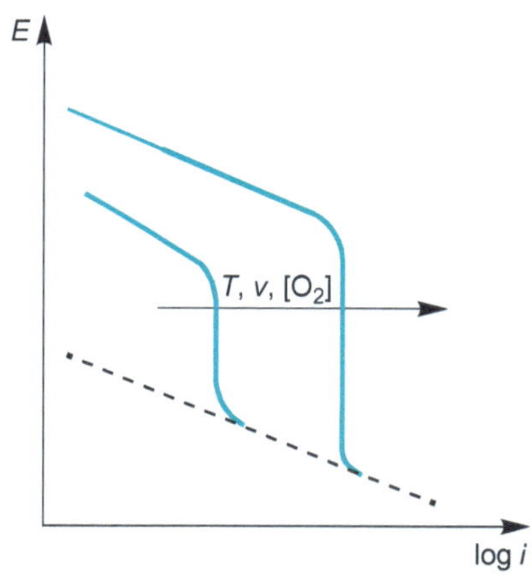

Figure 1.A2 Parameters influencing the oxygen limiting current density.

In order not to employ overly powerful feeders for passivation, which would be useless later for passivity maintenance, a common strategy is a progressive passivation operated by filling the tank step-by-step: for instance by adopting a feeder with maximum current output of 10 A, the tank can be filled to cover (i.e. to wet) $10/2.5 = 4\ m^2$, then once that surface is passivated, another equivalent filling step is performed until the full tank will be passivated.

1.7.2 Case study—design current for cathodic protection

Cathodic protection (CP) is an electrochemical technique to maintain fully or partially cathodic a metal exposed to an electrolyte, aiming to reduce or even halt corrosion.

Table 1.A2 Comparison between calculated and recommended values of protection current density for CP applications in seawater

Geographical location	$[O_2]$ content[a]	Temperature[a]	Velocity[a]	Calculated value	Value recommended by standards
	(ppm)	(°C)	(m/s)	(mA/m^2)	(mA/m^2)
North Sea	12 (oversaturation)	10	2	191	180–220
Mediterranean Sea; Gulf of Mexico; Indonesia; Persian Gulf	6	20	1	120	110
Australian Coasts; Brazil; West Africa	8	18	1.5	147	130
Sea mud (all seas)	3[b]	15	0	23	25

[a]Annual average.
[b]Overestimated.

CP is settled up by delivering a cathodic current to the exposed metal or, in other words, by supplying electrons. To stop corrosion, the number of electrons to be supplied, as minimum, is the same as those involved in the cathodic processes [3].

CP is successfully and widely adopted in natural environments, i.e. waters and soil, as well as in reinforced concrete structures, where cathodic process is oxygen reduction. In acids, where prevailing cathodic process is hydrogen evolution, CP is not used because protection current would be terribly high as well as the production of hydrogen gas.

The main design parameter is protection current, I_{prot}, given by the protection current density, i_{prot}, and the exposed metal surface, S: $I_{prot} = i_{prot} \cdot S$. The protection current density, i_{prot}, is the cathodic current density which coincides is the oxygen limiting current density, i_L.

Regardless the environment and the metal, the oxygen limiting current density is given by Eq. (1.35), where influencing parameters are temperature, T, oxygen content, $[O_2]$, and fluid velocity, v. As those parameters increase, oxygen limiting current density increases, as shown in Fig. 1.A2. By applying Eq. (1.35), protection current density can be easily estimated once influencing parameters are known. Table 1.A2 shows the comparison between calculated values and those specified by standards [5].

References

[1] Pedeferri P. Corrosione e protezione dei materiali metallici. Italy: Milan: Polipress; 2007 ISBN 97888-7398-032-2 (in Italian).
[2] Piontelli R. Elementi di teoria della corrosione a umido dei materiali metallici. Italy: Milan: Longanesi; 1961 (in Italian).
[3] Lazzari L, Pedeferri P. Cathodic Protection. Italy: Milan: Polipress; 2005 ISBN 88-7398-020-1.
[4] Pourbaix M. Lectures on electrochemical corrosion. New York: Plenum Press; 1973 ISBN 0-306-30449-X.
[5] EN Standard, EN 12495, Cathodic Protection for Fixed Steel Offshore Structures. European Committee for Standardization, B-1050. Belgium: Brussels; 2000.

Further reading

Fontana M. Corrosion engineering. 3rd Ed. New York: McGraw-Hill; 1986 ISBN 0-07-100360-6.
Bianchi G, Mazza F. Corrosione e protezione dei metalli. 3a Ed. Italy: Milan: Masson Italia Editori; 1989 ISBN 8-8214-0071-9 (in Italian).
Bardal E. Corrosion and protection. London: Springer-Verlag. London Ltd; 2004 ISBN 1-85233-758-3.

Uniform Corrosion

2

2.1 Model for acidic corrosion

Tafel equation lets to calculate, with good approximation, corrosion rate of metals, for instance ferrous alloys (carbon and low alloy steels) and zinc, in acidic solutions, when the predominant cathodic reaction is hydrogen evolution. Typical acidic solutions present in industry are as follows: strong acids (hydrochloric and sulphuric acids), carbonic acid (when carbon dioxide is present) hydrogen sulphide and organic acids.

The model here proposed is well illustrated by Evans diagram for active metals, such as iron and zinc, in acidic solutions where the cathodic process is hydrogen evolution. Fig. 2.1 shows an example of Evans diagram for two metals in an acidic solution with same pH (hence, same equilibrium potential of the hydrogen evolution reaction) and assuming anodic Tafel slope as near zero ($b_a \cong 0$) for the metal dissolution reaction, which means that the free-corrosion potential of the metal is taken as its equilibrium potential. This approximation is acceptable when cathodic reaction is hydrogen evolution, only, or largely prevalent (i.e. oxygen-free/chlorine-free acidic solutions). The equilibrium potential of the metal is given by the Nernst equations:

$$E_{eq,M} = E^0 + \frac{0.059}{2} \log a_{M^{2+}} \quad (2.1)$$

where M is the Fe or Zn or other active metals, for instance Cu in copper-complexant solutions. The cathodic curve, i.e. the Tafel straight line of hydrogen evolution, is given by Tafel equation (1.26) which can be rewritten as follows:

$$i = i_0 \cdot 10^{\frac{\eta}{b}} \quad (2.2)$$

where b is the module of Tafel slope for hydrogen evolution equals to 0.12 V/decade, i_0 is the exchange current density for hydrogen evolution on the metal (see Table 1.5); η is given by $E_{eq,H} - E_{corr}$, where E_{corr} is taken as $E_{eq,M}$, hence:

$$\eta = E_{eq,H} - E_{eq,M} = -0.059\ \text{pH} - \left(E^0 + \frac{0.059}{2}\left(\frac{T+273}{298}\right)\log a_{M^{2+}}\right) \quad (2.3)$$

Engineering Tools for Corrosion. DOI: http://dx.doi.org/10.1016/B978-0-08-102424-9.00002-1

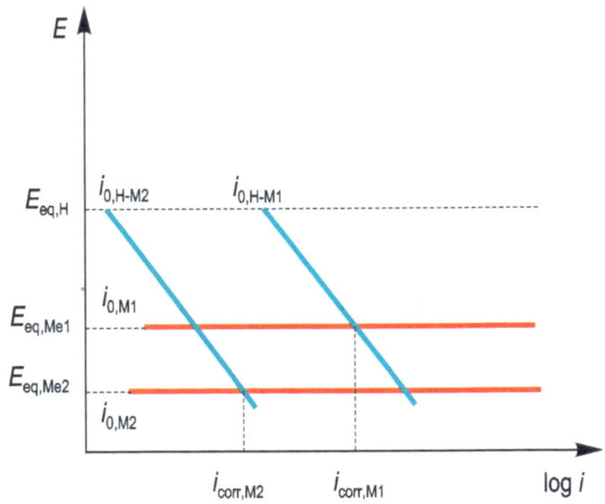

Figure 2.1 Evans diagram to determine corrosion rate in acidic solutions for active metals.

Introducing parametrically the fluid velocity with a parabolic dependence, as derived from literature data, and temperature as affecting equilibrium potential, the following empirical equation can be used:

$$i_{\text{corr,acid}} = i_{0,\text{H}} \cdot (1 + v^2) \cdot 10^{\frac{-0.059\text{pH} - \left(E^0 + \frac{0.059}{2}\left(\frac{T+273}{298}\right)\log a_{\text{M}^{2+}}\right)}{0.12}} \tag{2.4}$$

where required input parameters are:

- $i_{0,\text{H}}$ (exchange current density of hydrogen evolution on considered metal);
- v (fluid velocity in m/s);
- pH of the acid solution;
- $a_{\text{M}^{2+}}$ is metal ion concentration in the diffusion layer (interface) metal ion concentration at metal surface is derived from testing; in the case of formation of insoluble corrosion products or complex, concentration of metal ions is derived from solubility or complex constant;
- T is temperature in °C.

To use Eq. (2.4) for practical applications, there is the need to input metal ion concentration, $a_{\text{M}^{2+}}$, which is in practice the only unknown or uncertain variable. For different types of acid solutions, Table 2.1 reports the suggested metal ion concentration, $a_{\text{M}^{2+}}$.

2.1.1 Strong acids

For strong acid solutions, the concentration of metal ions, typically Zn, Al, Fe and ferrous alloys, is 10^{-2} M in stagnant conditions. It should be noted that in flowing conditions that concentration lowers, however, the effect of velocity of the solution is introduced parametrically in Eq. (2.4) because the estimation of metal ion

Table 2.1 Parameters for calculation of corrosion rate in different acidic solutions

Active metal or alloy	Exchange current density of hydrogen evolution on metal, M, $i_{0,H\text{-}M}$ (mA/m^2)	Metal ion M^{2+} concentration [$a_{M^{2+}}$] mol/L			
		Strong acids	Organic acids	Hydrogen sulphide (H_2S)	Carbonic acid (H_2CO_3)
Zn	10^{-3}				10^{-6}
Cu	1	10^{-2}	10^{-10}	10^{-9}	
Fe (ferrous alloys)	1				10^{-6}–10^{-9}[a]

[a]From ambient temperature to 80°C.

concentration in flowing conditions is uncertain. In general, the corrosion rate in strong acid can be estimated with the following equation:

$$i_{corr,strong\text{-}acids} = i_{0,H} \cdot (1 + v^2) \cdot 10^{\frac{-0.059pH - \left(E^0 - 0.059\left(\frac{T+273}{298}\right)\right)}{0.12}} \tag{2.5}$$

where symbols are known.

2.1.2 *Carbonic acid*

Carbon steel and low alloy steel corrode in carbonic acid-containing solutions as well known in oil & gas industry. Usual approach in industry is empirical and equations used to estimate the corrosion rate, starting from the very first one by de Waard and Milliams [1], are derived from laboratory testing. In the following, same results are obtained from Tafel equation (2.4) by inputting related effecting parameters. pH of a solution of carbonic acid is derived from the equilibrium partial pressure of CO_2, pCO_2, as follows:

$$pH = -\log\left[10^{-4} \cdot (pCO_2)^{0.5}\right] = 4 - 0.5 \cdot \log\ pCO_2 \tag{2.6}$$

where pCO_2 is expressed in bar. By introducing in (2.4), corrosion rate of carbon steel in carbonic acid, i_{corr,CO_2} (mA/m^2), is given by the following equation:

$$i_{corr,CO_2} = i_{0,H} \cdot \left(1 + v^2\right) \cdot 10^{\frac{0.03\ \log(pCO_2) - \left(E^0 + 0.0295\left(\frac{T+273}{298}\right)\log\ a_{M^{2+}}\right) - 0.236}{0.12}} \tag{2.7}$$

where parameters are known from operating data except metal ion concentration, $a_{M^{2+}}$ (mol/L), which is determined by the solubility constant of iron carbonate, which lowers as temperature increases. In appendix, results obtained by Eq. (2.7) and de Waard & Milliams base equation are compared. As matching is good and because de Waard & Milliams base equation was obtained from regression analysis of laboratory test results, Eq. (2.7) is indirectly validated.

2.1.3 Hydrogen sulphide

Hydrogen sulphide (H_2S) dissolves in water forming a weak acid solution with pH around 6. With this pH, corrosion rate of carbon steel would be low or even negligible; conversely, experience showed that corrosion rate is remarkable because insoluble FeS forms, having a solubility constant of 10^{-24}, which lowers equilibrium potential of iron. In Eq. (2.3), ion iron concentration would be $a_{Fe^{2+}} = 10^{-12}\ \text{mol/L}$, but experienced corrosion rates fit with an input of $a_{Fe^{2+}} = 10^{-9}\ \text{mol/L}$, as reported in Table 2.1, because most likely FeS layer works as barrier. This effect is similar to the one occurring on carbon steel at temperature above 120°C when exposed to carbonic acid because the iron carbonate becomes very protective.

By introducing in the general Eq. (2.4) pH = 6, $a_{Fe^{2+}} = 10^{-9}\ \text{mol/L}$, $i_{0,H} = 1\ \text{mA/m}^2$ and discarding the influence of fluid velocity, corrosion rate (μm/y) of carbon and low alloy steels is given by the following equation which is influenced by the temperature only, valid for temperature up to 60°C:

$$C_{\text{rate},H_2S} \cong 10^{0.66+2.2\cdot\left(\frac{T+273}{298}\right)} \tag{2.8}$$

The resulting maximum corrosion rate is about 1 mm/y, as experience has confirmed. A more general empirical equation can be derived by introducing a parabolic dependence as follows:

$$C_{\text{rate},H_2S} \cong 10^3 \cdot \left[1 - \left(\frac{T-60}{60}\right)^2\right] \tag{2.9}$$

where corrosion rate is in μm/y and temperature in °C. Above 60°C, corrosion rate decreases progressively because, although its electronic conductance, iron sulphide has strong protection properties based on barrier effect. Eq. (2.9) is valid up to 120°C.

2.1.4 Organic acids

For complexant organic acids, the model is still the same where metal ion concentration is derived from the complex constant. The value of the complex constant varies with the acid and metal; therefore, the general Eq. (2.4) is applicable by introducing pH and the metal ion concentration at the equilibrium with the complex.

However, since complex constant is often not known, the order of magnitude of expected corrosion rate is obtained by introducing in the general Eq. (2.4) a metal ion concentration of $a_{Fe^{2+}} = 10^{-10} - 10^{-8}\ \text{mol/L}$ and an exchange current density for hydrogen evolution on carbon steel and stainless steels of $i_{0,H} = 1$ mA/m^2. At pH of about 5 to 6, corrosion rates vary from 0.5 to 12 mm/y for a temperature range between 25 and 100°C. For acetic acid, see appendix.

Some corrosion attacks attributed to MIC (see Chapter 5: Corrosion in Water, Soil and Air, par. 5.5) are caused by the presence of organic acids released by bacteria.

2.1.5 Uniform corrosion of passive metals in acids

In acidic solutions, passive metals can show:

- depassivation (i.e. oxide dissolution) then behaving as an active metal so following the model summarized by Fig. 2.1;
- passivity without risk of localized corrosion (i.e. chloride content is below the critical threshold content and free-corrosion potential is below pitting potential);
- passivity with potential risk of localized corrosion (i.e. chloride content is above the critical threshold content but free-corrosion potential is below pitting potential).

Uniform corrosion may occur if depassivation takes place. This occurrence is possible if pH is below depassivation pH, which, for low grade stainless steels is about 2 and much lower for nickel-based alloys. Uniform corrosion rate can be predicted by the model discussed above.

If there is no depassivation, as summarized in Fig. 2.2, two conditions exist:

- *Absence* of chlorides. Passivity current density of most common passive metals (i.e. stainless steels, nickel-based alloys, titanium alloys) is less than exchange current density of hydrogen evolution, therefore free-corrosion potential is equilibrium potential of hydrogen evolution (-0.059pH) and driving voltage for hydrogen evolution is zero;
- *Presence* of chlorides. As of the presence of chlorides, passivity current density may exceed the exchange current density of hydrogen evolution, leading to a free-corrosion potential lower than equilibrium potential of hydrogen revolution (-0.059pH), with a corrosion rate that is the passivity current density given by Tafel equation:

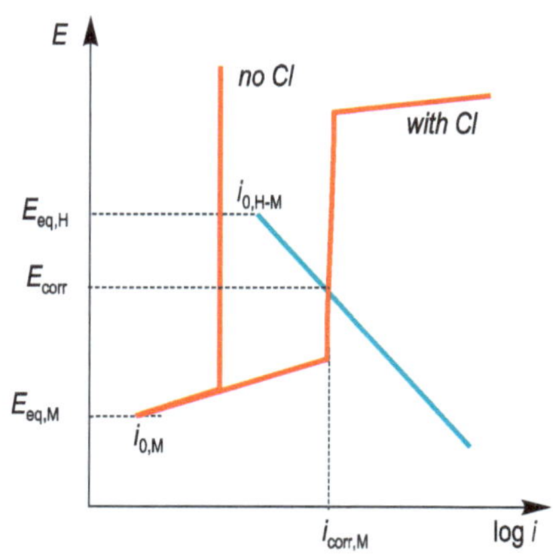

Figure 2.2 Evans diagram to determine corrosion rate in acidic solutions of passive metals.

$$i_{corr} = i_p = i_{cath,H} = i_{0,H} \cdot 10^{\frac{-0.059pH - E_{corr}}{0.12}} \tag{2.10}$$

where $i_{0,H}$ is the exchange current density of hydrogen evolution on considered metal, see Table 1.5, E_{corr} is free-corrosion potential.

As passivity current density in the presence of chlorides is in the same range of exchange current density of hydrogen evolution, free-corrosion potential is as a maximum about 100 mV more negative than hydrogen equilibrium potential.

2.2 Aerated solutions

In aerated near neutral solutions, the cathodic processes are: oxygen reduction first and hydrogen evolution when potential drops below the equilibrium potential, according to the metal involved. For instance, in aerated acid solutions, if metal is noble like copper or silver, the only possible cathodic reaction is oxygen reduction (plus chlorine reduction if present). If metal is iron and pH exceeds neutrality, again the only practical cathodic process is oxygen reduction (plus chlorine reduction if present); if metal is more electronegative such as zinc, both reaction, i.e. oxygen reduction and hydrogen evolution, take place.

2.2.1 Oxygen limiting current density

When oxygen reduction reaction takes place below the field of activation overvoltage, diffusion controls the oxygen availability through the oxygen limiting diffusion current density, i_L, which is governed by Fick Law on diffusion, as seen in Chapter 1, Basic Principles:

$$i_L \cong 10 \cdot 2^{\frac{T-25}{25}} \cdot [O_2] \cdot (1 + \sqrt{v}) \tag{2.11}$$

where i_L is in mA/m^2; $[O_2]$ (mg/L $\approx$ ppm) is the oxygen content in water; v (m/s) is the water velocity, and T (°C) is the temperature.

2.2.2 Presence of chlorine

The presence of chlorine gives rise a more noble cathodic process, which takes place first:

$$Cl_2 + 2e \rightarrow 2Cl^- \tag{2.12}$$

Chlorine, like oxygen, is a gas that dissolves in water, but unlike oxygen it partially dismutises so the fraction available for diffusion is about 30% (chlorine diffusion coefficient is 1.38 10^{-9} m^2/s at 25°C and like oxygen it doubles every about 25°C). According to this, Eq. (2.11) can be revised by introducing also chlorine taking into account the different diffusion coefficient and valence (8 g of oxygen is equivalent to 35 g of chlorine); eventually, it becomes

$$i_L \cong 10 \cdot 2^{\frac{T-25}{25}} \cdot \{[O_2] + 0.04 \cdot [Cl_2]\} \cdot (1 + \sqrt{v}) \tag{2.13}$$

where i_L is in mA/m^2; $[O_2]$ and $[Cl_2]$ (mg/L $\approx$ ppm) are oxygen and chlorine concentrations in water, respectively; v (m/s) is the water velocity, and T (°C) is the temperature.

2.2.3 Dimensionless number approach

In flowing condition, diffusion layer thickness, δ, can be calculated by classic hydrodynamic approach based on Sherwood (or Nusselt) dimensionless number, which gives:

$$Sh = \frac{\phi}{\delta} = i_L \frac{\phi}{4FD[O_2]} \tag{2.14}$$

where ϕ (m) is called characteristic dimension, for example the pipe diameter; the meaning of other parameters is known. Sherwood number is a function of Reynolds (Re) and Schmidt (Sc) dimensionless numbers, which are as follows:

$$Sh = 0.023 \cdot Re^{0.87} \cdot Sc^{0.33} \qquad Re = \frac{\phi v}{\upsilon} \quad Sc = \frac{\upsilon}{D} \tag{2.15}$$

where ϕ (m) is the characteristic dimension; v (m/s) is the water velocity; υ (m^2/s) is the cinematic viscosity and D (m^2/s) is the diffusion coefficient. Oxygen limiting current density, i_L, (A/m^2) is given by Sherwood number, which is as follows:

$$i_L = 4FD[O_2]\frac{Sh}{\phi} \tag{2.16}$$

where symbols are known. Diffusion coefficient, D, in (cm^2/s) is given by the following empirical equation:

$$\log D = -4.410 + 773.8/(T + 273) - [506.4/(T+273)]^2 \tag{2.17}$$

where T (C°) is the temperature.

2.3 Summary

Table 2.2 summarizes equations used for the calculation of uniform corrosion rates in acidic and aerated solutions.

Table 2.2 Equations for the calculation of uniform corrosion rates in acids and aerated solutions

	Equations	Parameters
Strong acids	$i_c = i_{0,H} \cdot 10^{\frac{\eta}{b}}$ $i_{corr,acid} = i_{0,H} \cdot (1+v^2) \cdot 10^{\frac{-0.059pH - \left(E^0 + \frac{0.059}{2}\left(\frac{T+273}{298}\right)\log a_{M^{2+}}\right)}{0.12}}$ $i_{corr,strong\text{-}acids} = i_{0,H} \cdot (1+v^2) \cdot 10^{\frac{-0.059pH - \left(E^0 - 0.059\left(\frac{T+273}{298}\right)\right)}{0.12}}$	$a_{M^{z+}}$ (mol/L) metal ion concentration b (V/decade) Tafel slope $[Cl_2]$ (ppm) chlorine content C_{rate} (μm/y) corrosion rate D (m^2/s) diffusion coeff E^0 (V) standard potential E_{corr} (V) free-corrosion potential η (V) overvoltage F (C) Faraday constant ϕ (m) characteristic size i_c (A/m^2) cathodic current density i_L (A/m^2) oxygen limiting current density $i_{0,H}$ (A/m^2) exchange c.d. for hydrogen evolution on metal; $i_{0,H} = 1$ mA/m^2 for iron i_{corr} (mA/m^2) corrosion rate υ (m^2/s) cinematic viscosity $[O_2]$ (ppm) or (mol/m^3) oxygen content pCO_2 (bar) partial pressure of carbon dioxide T (°C) temperature v (m/s) fluid velocity
Carbon steel in carbonic acid	$i_{corr,CO_2} = i_{0,H} \cdot (1+v^2) \cdot 10^{\frac{0.03\log(pCO_2) - \left(E^0 + 0.0295\left(\frac{T+273}{298}\right)\log a_{M^{2+}}\right) - 0.236}{0.12}}$ $pH = 4 - 0.5 \cdot \log pCO_2$	
Carbon steel in hydrogen sulphide	$C_{rate,H_2S} \cong 10^{0.66+2.2\cdot\left(\frac{T+273}{298}\right)}$ $C_{rate,H_2S} \cong 10^3 \cdot \left[1 - \left(\frac{T-60}{60}\right)^2\right]$	
Stainless steel in strong acids	$i_{corr} = i_p = i_{cath,H} = i_{0,H} \cdot 10^{\frac{-0.059pH - E_{corr}}{0.12}}$	
Diffusion limiting current density (oxygen + chlorine) (mA/m^2)	$i_L \cong 10 \cdot 2^{\frac{T-25}{25}} \cdot [O_2] + 0.04 \cdot [Cl_2] \cdot (1+\sqrt{v})$	
Oxygen limiting current density (mA/m^2)	$i_L = 4FD[O_2]\frac{Sh}{\phi}$ $Sh = 0.023 \cdot Re^{0.87} \cdot Sc^{0.33}; \quad Re = \frac{\phi v}{\upsilon}; \quad Sc = \frac{\upsilon}{D}$ $\log D\left[cm^2\ s^{-1}\right] = -4.410 + 773.8/(T+273) - [506.4/(T+273)]^2$	

2.4 Appendix

2.4.1 Coefficient of variation, CV

As discussed in Chapter 8, Statistical Analysis of Corrosion Data, corrosion rates calculated by the algorithms have a stochastic meaning, because in practice a corrosion process occurs through a distribution of values rather than a fixed, constant, determined one.

The use of the calculated value, as well as for measured values in testing or in inspections, should take into account this interpretation. As said, CV derived from the experience can be suitably used for further calculations such as the maximum expected value.

For uniform corrosion in acid and aerated solutions, the following CV values can be adopted:

- Acidic solutions: CV = 0.2 (in some instance, 0.4 could also be considered). The maximum expected value can be taken as $\mu + 2\sigma$, where the standard deviation, σ, is obtained from the coefficient of variation, CV. By assuming a CV equals to 0.2, the maximum value is $x_{max} \cong 1.4\ \mu$;
- Aerated solutions: CV = 0.4 (in some instance, 0.6 could also be considered). The maximum expected value can be taken as $\mu + 2\sigma$, where the standard deviation, σ, is obtained from the coefficient of variation, CV. By assuming a CV equals to 0.4, the maximum value is $x_{max} \cong 1.8\ \mu$. By applying the Gumbel statistics, $x_{max} = 2.25\ \mu$.

2.4.2 Corrosion rate in carbonic acid

In industrial applications, corrosion rate of carbon steel exposed to carbonic acid (i.e. in carbon dioxide containing hydrocarbons) is calculated by the de Waard & Milliams equation, which is derived from laboratory testing. The base equation is the following [1]:

$$\log C_{rate,dWM} = 5.8 - \frac{1710}{T + 273} + 0.67 \cdot \log pCO_2 \tag{2.18}$$

where $C_{rate,dWM}$ is the corrosion rate (mm/y); T is the temperature (°C), and pCO_2 is the carbon dioxide partial pressure (bar). Some empirical coefficients are used to either mitigate or increase the base calculated corrosion rate when temperature exceeds so-called scaling temperature and pH is lower than the equilibrium value [2].

For comparison, Table 2.A1 reports corrosion rates calculated by the model and de Waard & Milliams base Eq. (2.18). Match is good at low temperature; instead, it seems that Eq. (2.7) does not need any correction at high pressure and high CO_2 partial pressure.

Table 2.A1 Comparison between corrosion rates of carbon steel in carbonic acid calculated by the model and by de Waard & Milliams base equation

		$pCO_2 = 1$ bar (mm/y)		$pCO_2 = 2$ bar (mm/y)		$pCO_2 = 5$ bar (mm/y)		$pCO_2 = 10$ bar (mm/y)	
T (°C)	$a_{M^{2+}}$ (mol/l)	Model	deW&M	Model	deW&M	Model	deW&M	Model	deW&M
25	10^{-6}	1.7	1.2	2.1	1.8	2.6	3.4	3.1	5.4
50	10^{-7}	4.3	3.2	5.1	5.1	6.4	9.4	7.6	15
75	10^{-8}	11.6	7.7	13.8	12.2	17.4	22.6	20.6	36
100[a]	$10^{-8.5}$	24.2	16.4	28.8	26.1	36.2	48	43	76.8

[a]Real corrosion rates are much lower because a protective carbonate, $FeCO_3$, forms.

2.4.2.1 *Corrosion mechanism in carbonic acid*

There is a variety of references on this matter. In short, the most cited and accepted mechanism for corrosion of carbon steels in carbonic acid is, almost arbitrarily, the following:

- because corrosion rate in carbonic acid, although it is a weak acid, is an order of magnitude higher than the one in strong acid at the same pH (3 to 6), there should be more than one cathodic reaction besides the hydronium reduction;
- other cathodic reactions would be the direct reduction of either bicarbonate ion or undissociated carbonic acid:
  - $H_2CO_3 + 2e = H_2 + CO_3^{2-}$;
  - $2HCO_3^- + 2e = H_2 + 2CO_3^{2-}$.

As shown in Table 2.A1, same corrosion rates as from testing (from which de Waard & Milliams equation was derived) are obtained on the basis of the corrosion theory:

- the cathodic reaction is hydrogen evolution (to give hydrogen gas) following Tafel Law with 120 mV/decade as slope;
- free-corrosion potential approximates iron equilibrium potential, governed by the solubility product of $FeCO_3$ which is at room temperature $2 \cdot 10^{-11}$ and decreases strongly as temperature increases.

In conclusion, it seems that there is no need to claim exotic mechanisms or empirical model derived from testing. Indeed, results obtained from experimental tests confirm those calculated by a theoretical model, valid for all acid classes, i.e. strong, weak, organic acids, hydrogen sulphide and carbonic acid.

2.4.3 *Corrosion rate by organic acids*

Corrosion rates of metals exposed to organic acids are surprisingly high as reported in literature, for example [3–5]. The model proposed gives same order of magnitude of measured corrosion rates, as Table 2.A2 shows. For acetic acid, metal ion

Table 2.A2 Comparison of corrosion rate (mm/y) in deaerated organic acids: calculated vs test results [3,4]

		Mild steel				Stainless steel (304)			
		Acetic acid		Propionic acid		Acetic acid		Propionic acid	
T (°C)	Conc (%)	Calculated	Meas	Calculated	Meas	Calculated	Meas	Calculated	Meas
25	20	0.9 (pH = 3.5; $a = 10^{-4}$)	0.2			0.4 (pH = 3.5; $a = 10^{-4}$) <0.001 aerated	0.45 0.002 aerated		
25	100	1.9 (pH = 2.8; $a = 10^{-4}$)	0.8–1.7						
>100	5	15 (pH = 2.8; $a = 10^{-6}$)	50						
>100	100			33 (pH = 4; $a = 10^{-9}$)	34	1.8 (pH = 4; $a = 10^{-6}$) <0.02 aerated	1.2 0.002 aerated	1.8 (pH = 4; $a = 10^{-6}$) <0.02 aerated	0.6

concentration is of the order of magnitude of 10^{-4} at low temperature and of 10^{-6} at boiling temperature. For propionic acid, at boiling temperature, metal ion concentration is of orders of magnitude lower, from 10^{-6} for stainless steel to 10^{-9} for iron.

2.4.4 Corrosion rate in acidic solutions of Fe, Zn and Cu

Table 2.A3 reports the calculated corrosion rates by using the proposed model for typical acidic environments in industry.

2.4.5 Case study: stainless steel in acetic acid

This is a useful exercise for understanding how operating conditions affect the corrosion behaviour. Two conditions apply: aerated acetic acid and deaerated acetic acid as shown in Fig. 2.A1.

Corrosion condition in aerated acetic acid, where stainless steel resists corrosion because passive. Corrosion rate equals the passivity current density, in accordance with Eq. (1.24), hence $i_{corr} = i_p \cong 0.1$ mA/m^2 (by inputting: [Cl] = 0; PREN = 18; pH = 3; T = 25°C).

Table 2.A3 Corrosion rates for ferrous alloys and zinc in strong acids and copper in hot concentrated HCl

	T **(°C)**	C_{rate} **(mm/y)**		
		Fe	**Zn**	**Cu**
E^0 (V SHE)		−0.44	−0.76	0.34
[Me^{2+}] (mol/L)		0.01	0.01	10^{-14}
$i_{0,H}$ (mA/m^2)		1	0.001	1
pH = −1	25	52	21	0.01
	50	57	24	0.02
	100	69	29	0.06
pH = 4	25	0.18	0.06	
	50	0.20	0.08	
	100	0.24	0.09	
pH = 6	25	0.019	0.008	
	50	0.021	0.009	
	100	0.025	0.01	

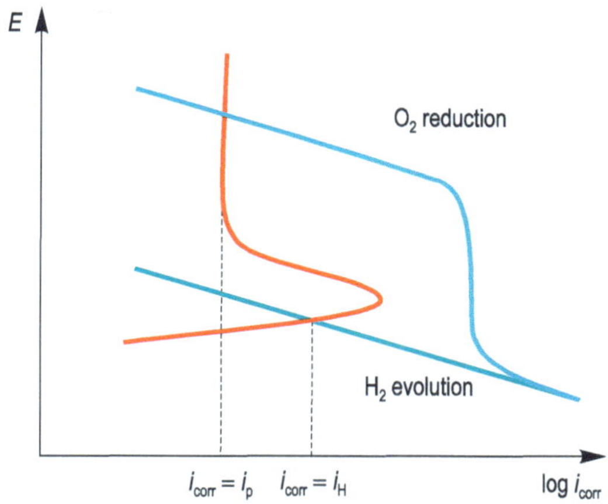

Figure 2.A1 Corrosion condition of AISI 304 SS in aerated and deaerated acetic acid.

Corrosion condition in deaerated acetic acid, where stainless steel cannot passivate. Mean corrosion rate is given by Eq. (2.5), hence $i_{corr,acid} \cong 0.7$ mA/m^2 [by inputting: $v = 0$ (m/s); pH = 3; $i_{0,H} = 1$ mA/m^2; $E^0 = -0.4$ V (for stainless steel is a weighted average); $[M^{z+}] = 10^{-4}$ mol/L (as minimum); $T = 25°C$]. Corrosion attack is uniform; hence, max expected value is $\mu + 2\sigma$; assuming a CV = 0.2, $C_{rate,max} \cong 1$ mm/y. These values fit with reference data as reported in [6].

2.4.6 *Case study: stainless steel in hot acids*

A plant for treating a hydrochloric acid-containing exhaust gas consists of a neutralizing tower where a sodium hydroxide solution is injected through squirrels to neutralize the hydrochloric acid. First choice of material is duplex stainless steel for a twofold reason: first, resisting chloride-induced stress corrosion cracking (SCC), and second, resisting acid corrosion before neutralization. Indeed, as far as SCC resistance is concerned, duplex is a correct choice; about the second requisite, there are two corrosion conditions to resist with: carbonic acid (which is affordable) and concentrated HCl solution, about 1 mol, to which corrosion rate should be checked.

To confirm the material selection, a corrosion assessment is mandatory. Gas composition (by volume) is approximately: 8% O_2, 0.3% HCl, 10% CO_2 and 16%H_2O.

Corrosion condition in hot dry gas. At low temperature, but above dew point, hot gas is not corrosive because no water condensates, hence no acidic solution forms.

Corrosion condition in hot wet gas. Below dew point, water condensates then producing an acidic solution about 1 mol of HCl and containing also carbonic acid (CO_2 partial pressure about 10%). After the injection of neutralizing solution, condensate is alkaline, with a pH close to 10.

To check the former condition, expected corrosion rate can be estimated by applying Eq. (2.5) because stainless steel depassivates. By inputting: $T \cong 100°C$; pH $\cong 0$; fluid velocity $\cong 2$ m/s; $i_{0,H} = 1$ mA/m^2; $E^0 = -0.4$ V SCE, $C_{rate} \cong 1.2 \times 43 = 52$ mm/y (or about 1 mm a week!).

Corrosion condition in alkaline solution. In neutral to alkaline solution, duplex stainless steel is passive as proved by the calculated passivity current density by Eq. (1.25), which is about 0.02 mA/m^2 (inputs are [Cl] $\cong$ 30,000 ppm; PREN 40; pH = 10; T = 100°C).

2.4.7 Dimensionless number approach vs empirical Fick equation

It could be of interest the comparison of results obtained by the "Dimensionless Number Approach", Eq. (2.16) and the Fick Law, Eq. (2.11). For instance, by assuming the following typical input data:

- oxygen content: 10 ppm (mg/L) = 0.3 mol/m^3;
- room temperature;
- water velocity: 1, 2 and 3 m/s;
- cinematic viscosity: 0.001 m^2/s;
- diffusion coefficient: $2 \cdot 10^{-9}$ m^2/s;
- tube size: 0.5 and 1 m.

Results show a maximum scattering of $\pm 20\%$ and a good match for following case studies:

- v = 2 m/s, diameter 1 m, i_L = 241 vs 212 mA/m^2;
- v = 2 m/s, diameter 0.5 m, i_L = 241 vs 232 mA/m^2.

References

[1] De Waard C, Milliams DE. Carbonic acid corrosion of steel. Corrosion 1975;31(5).
[2] EFC Publication Number 23, *CO2 Corrosion Control in Oil and Gas Production*: Design *Considerations*, Ed. By Kermani M.B., Smith L.M., The Institute of Materials, 1997.
[3] Corrosion by acetic acid, A report task group T-5A-3, Pub. 57-25, NACE; TX; Houston; Nov. 1957, pp. 757t–766t.
[4] Corrosion of metals by aliphatic organic acids, A report task group T-5A-6c, Pub. 5A 180, *Mat. Perform.*, NACE, Houston, Sept. 1980, TX, pp. 65–68.
[5] *Corrosion of metals by acetic acid*, Dechema, Vol. 6, pp. 44–49.
[6] *Corrosion data survey, Metal section*, VI edition, NACE Int.; TX; Houston; 1985. ISBN 0-915567-07-5.

Further reading

Pedeferri P. Corrosione e protezione dei materiali metallici. Milan, Italy: Polipress; 2007 ISBN 97888-7398-032-2 (in Italian).

Localized Corrosion

3

3.1 Macrocell

Localized corrosion propagates by means of so-called macrocell mechanism (Fig. 3.1):

- corroding zone (often of small surface area) is the anode;
- an associated area hosts the cathodic process (in general oxygen reduction);
- a current flows from anodic zone to the cathodic ones, determining the corrosion rate.

Unlike galvanic corrosion, on each anodic and cathodic zones, only anodic and cathodic reactions take place, respectively. Hence, corrosion rate by macrocell mechanism, $C_{\mathrm{rate,MC}}$, is given by the following equation:

$$C_{\mathrm{rate,MC}} = k \cdot i_{\mathrm{C}} \cdot \frac{S_{\mathrm{c}}}{S_{\mathrm{a}}} \tag{3.1}$$

where $C_{\mathrm{rate,MC}}$ is in mm/y if current density of the cathodic reaction, i_{C}, is in A/m^2; S_{C} and S_{A} are anodic and cathodic surface area, respectively, and k is the constant equivalence (see Table 1.1; for carbon steel $k = 1.2$). Surface area ratio, $S_{\mathrm{C}}/S_{\mathrm{A}}$, is not always the one we think or simply we measure, because it is determined by the electrical field setup by the macrocell. In other words, it is determined by the so-called *throwing power*, L_{max}. For instance, for a plate-like geometry of Fig. 3.2, cathodic surface area, S_{C}, is $\pi \cdot (L_{\mathrm{max}})^2$, which is limited to a portion of the plate, only.

When a macrocell sets up, there is a potential distribution on the exposed metal surface (so-called potential mapping), which also extends its influence within the electrolyte. Fig. 3.3 shows how potential qualitatively changes as distance from the

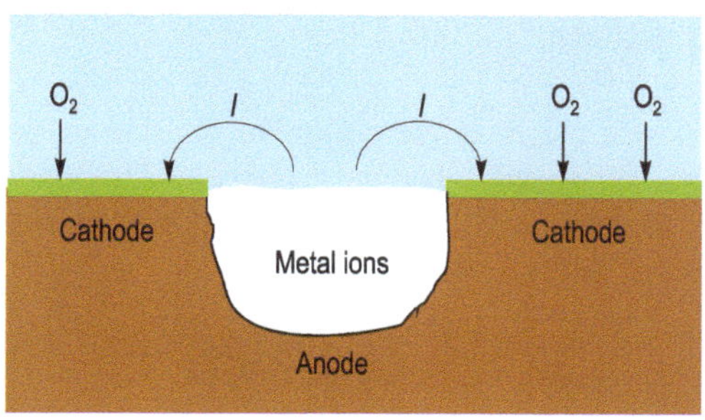

Figure 3.1 Schematic representation of a macrocell corrosion mechanism (cathodic process is oxygen reduction).

Engineering Tools for Corrosion. DOI: http://dx.doi.org/10.1016/B978-0-08-102424-9.00003-3

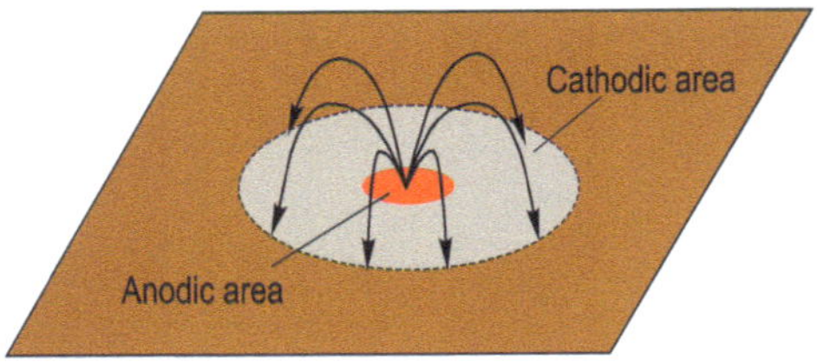

Figure 3.2 Throwing power on a plate-like cathodic area for a localized corrosion proceeding by macrocell mechanism.

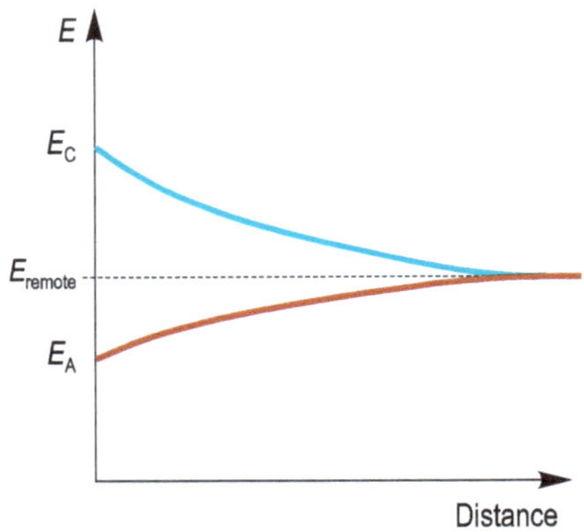

Figure 3.3 Potential trend with distance from a metal surface where a macrocell works.

metal surface increases; beyond a certain distance from the metal surface, measured potential is constant and is called *remote potential*.

3.2 Throwing power

The term throwing power refers to the ability of the current to reach areas distant from the anodic zone. It depends on both primary and secondary current distributions (see Chapter 6: Computer Modelling), i.e. through overvoltage, resistivity and geometry of the domain, as follows:

- driving voltage, ΔV (V), determined by the overvoltage. The driving voltage is dissipated in the electrolyte as ohmic drop;
- current density of the cathodic reaction, in most cases oxygen limiting current density, i_{lim} (A/m^2); it is considered as uniform on the cathodic surface;
- electrolyte resistivity (Ωm);
- geometry of the domain, which concurs to determine the ohmic drop in the electrolyte.

With reference to a simple cell geometry, as depicted in Fig. 3.1, the driving voltage, ΔV, dissipated as ohmic drop in the electrolyte, is given by a general expression as follows:

$$\Delta V = I \cdot R = \rho \int_{n} \frac{i_n}{S} dL \cong k \cdot \rho \cdot i \cdot L \tag{3.2}$$

where k is the cell constant, which depends on electrolyte geometry; L is the distance between the two electrodes; i is an average current density within the electrolyte. Throwing power, L_{max}, for simplified geometry for which relationship is linear, is expressed by an approximate expression, which is as follows:

$$L_{max} \cong k \frac{\Delta V}{\rho \cdot i_C} \tag{3.3}$$

where current density is taken as the cathodic one. More generally as derived from a dimensional analysis approach:

$$L_{max} = k \sqrt{\frac{\Delta V \ \phi_k}{\rho \ i_C}} \tag{3.4}$$

where ϕ_k is the characteristic dimension; k is an appropriate constant for each geometry. Eq. (3.4) states that L_{max} does not vary linearly with driving voltage and

Table 3.1 Throwing power for typical macrocell geometries ($S_C \gg S_A$)

Geometry	Throwing power	
	Bare cathodic surface	**Coated cathodic surface**
Plate-like geometry (typical for pitting)	$L_{max} \cong \frac{1}{2} \frac{\Delta V}{(\rho + 0.8)(i + 10)}$ $L_{max} \cong \sqrt{2} \sqrt{\frac{\Delta V \ \phi_{anode}}{(\rho + 0.8)(i + 10)}}$	$L_{max} \cong \frac{1}{2} \frac{\Delta V}{(\rho + 0.8)(i + 0.1)}$ $L_{max} \cong \sqrt{2} \sqrt{\frac{\Delta V \ \phi_{anode}}{(\rho + 0.8)(i + 0.1)}}$
Inside a tube or a fissure	$L_{max} \cong \frac{1}{\sqrt{2}} \sqrt{\frac{\Delta V \ \phi_{tube}}{(\rho + 0.8)(i + 10)}}$	$L_{max} \cong \frac{1}{\sqrt{2}} \sqrt{\frac{\Delta V \ \phi_{tube}}{(\rho + 0.8)(i + 0.1)}}$
Outside a pipeline	$L_{max} \cong \frac{1}{2} \frac{\Delta V}{(\rho + 0.8)(i + 10)}$ $L_{max} \cong 10 \sqrt{\frac{\Delta V \ \phi_{pipe}}{(\rho + 0.8)(i + 10)}}$	$L_{max} \cong \frac{1}{2} \frac{\Delta V}{(\rho + 0.8)(i + 0.1)}$ $L_{max} \cong 10 \sqrt{\frac{\Delta V \ \phi_{pipe}}{(\rho + 0.8)(i + 0.1)}}$

L_{max}: throwing power (m); ϕ_k: characteristic size (m); i: cathodic current density (mA/m^2); ρ: resistivity (Ωm); ΔV: driving voltage dissipated as ohmic drop in the electrolyte (mV).

flowing current as confirmed by experience and FEM simulations. Table 3.1 reports empirical equations, validated by FEM simulations, for the calculation of throwing power for typical geometries, for bare and coated surfaces [1].

3.2.1 *Effective driving voltage, ΔV*

Driving voltage, ΔV, is formally given by the following equation:

$$\Delta V = (E_{eq,C} - \eta_C) - (E_{eq,A} + \eta_A) = IR_{el} \tag{3.5}$$

where $E_{eq,C}$ is the equilibrium potential of cathodic process, $E_{eq,A}$ is the equilibrium potential of anodic process, η_C and η_A are modules of cathodic and anodic overvoltage, respectively, IR_{el} is the resulting ohmic drop in the electrolyte.

It must be noted that driving voltage, $\Delta V \equiv IR_{el}$, is time-dependent as the circulating current varies with time. This is typical in galvanic corrosion and cathodic protection by galvanic anodes as considered in Section 3.4.4.

3.3 Surface area ratio

In a macrocell, throwing power defines inherently the maximum surface area ratio. As experience has proved, parameters determining the throwing power do affect also the surface area ratio setting up by the macrocell. Accordingly, a similar relationship used for throwing power applies for maximum surface area ratio as follows:

$$\left(\frac{S_C}{S_A}\right)_{max} \cong \left(1 + \frac{S_C}{S_A}\right)_{max} \cong k\sqrt{\frac{\Delta V}{i_C \cdot (\rho + 0.8)}} \tag{3.6}$$

where ΔV is the driving voltage (V); ρ is the resistivity (Ωm); i_C is the cathodic current density (A/m^2), and k is the dimensional constant (m$^{0.5}$). In near neutral-to-alkaline electrolytes, where cathodic process is oxygen reduction, cathodic current density is oxygen limiting current density, i_L, and constant k is about 20. Empirical equation (3.6) is useful to have a reliable reference, at least as order of magnitude; for a more accurate evaluation of surface area ratio, FEM simulation is recommended.

3.4 Galvanic corrosion

Galvanic corrosion, also called *bimetallic corrosion*, is produced when two metals immersed in an electrolyte are in electrical contact and are characterized by a different practical nobility, i.e. a different free corrosion potential. To have a galvanic attack, a cathodic process has to be possible, so that less noble metal, A, corrodes at an accelerated rate, whereas more noble metal, C, exhibits a decreased corrosion rate up to a halt. Materials with electronic conductivity can work as cathode, such as magnetite that forms near welds, calamine in hot lamination, magnetite in boilers and also graphite and sulphides in industrial processes.

By introducing surface areas, S_A and S_C, macrocell current, I, called galvanic current, is as follows:

$$I = I_A = I_C; \quad I_A = i_{corr} S_A; \quad I_C = i_C (S_A + S_C);$$
$$i_{corr} = i_C \cdot \frac{S_C + S_A}{S_A} \tag{3.7}$$

where i_{corr} is the corrosion rate of less noble metal, A, and i_C is the cathodic current density on *both metals*. When $S_C \gg S_A$, Eq. (3.7) approximates Eq. (3.1). The maximum cathodic-to-anodic surface area ratio is given by Eq. (3.6), therefore, by matching Eqs (3.1) and (3.7), maximum galvanic corrosion rate, for $S_C \gg S_A$, typically for plate-like geometry, is expressed by the following equation:

$$C_{rate,GC} \cong k \sqrt{\frac{i_C \cdot \Delta V}{(\rho + 0.8)}} \tag{3.8}$$

where symbols are known and $C_{rate,GC}$ is in mm/y. The dimensional constant, k, is determined experimentally or by FEM simulations; for active metals, such constant is about $k \cong 20$ ($m^{0.5}$), except for Pb and Sn for which it doubles. As summarized by Eq. (3.8), three parameters determine the corrosion rate in a galvanic coupling:

- driving voltage, as ohmic drop dissipation in the electrolyte;
- electrolyte resistivity;
- cathodic current density.

3.4.1 Driving voltage for galvanic corrosion

When considering a simplified model of macrocell, it is assumed that there is an anode potential, E_A, and a cathode potential, E_C. Truly, especially on the cathodic surface, there is a potential distribution, which becomes in practice flat when the cathodic process is oxygen reduction. When cathodic reaction is hydrogen evolution, the current distribution on the cathodic surface is far from a near-uniform one, as easily verified by FEM modelling, however, as an acceptable approximation, an average current density can be taken into consideration even in this case. According to this model, driving voltage, ΔV, is given by $E_C - E_A$.

3.4.1.1 Anode potential, E_A

As galvanic corrosion affects less noble active metals, the working potential of the anode is close to the equilibrium potential, which is easily calculated by the Nernst Equation:

$$E_{eq} = E^0 + \frac{0.059}{z}\log a_{M^{z+}} \tag{3.9}$$

Metal ion concentration $a_{M^{z+}}$ is often taken as 10^{-6} mol/L, as suggested by Pourbaix, for electrolytes not containing metal ions, as in the case of metals exposed to waters or immersed in soil. In stagnant condition and especially in strong acids, $a_{M^{z+}}$ assumes much higher concentration, as 10^{-2} mol/L and more.

It should be noted that by taking the potential of the anodic reaction as the equilibrium potential of the corroding metal, it is assumed that anodic overvoltage is negligible.

3.4.1.2 Cathode potential, E_C, for oxygen reduction

Figs 3.4 and 3.5 depict the electrochemical conditions for two typical galvanic corrosion conditions in neutral, aerated solutions: one refers to an active cathodic metal and the second one to a passive cathodic metal.

Fig. 3.4A shows the case of cathodic active metal, for instance copper as E_{C1} against zinc and iron E_{C2} against zinc. The potential of the cathode is the equilibrium potential of the metal calculated by Eq. (3.9). Fig. 3.4B shows the case of cathodic noble metal as gold or platinum, for which the potential of the cathode is the equilibrium potential of oxygen (see Section 1.4.3.2), given by:

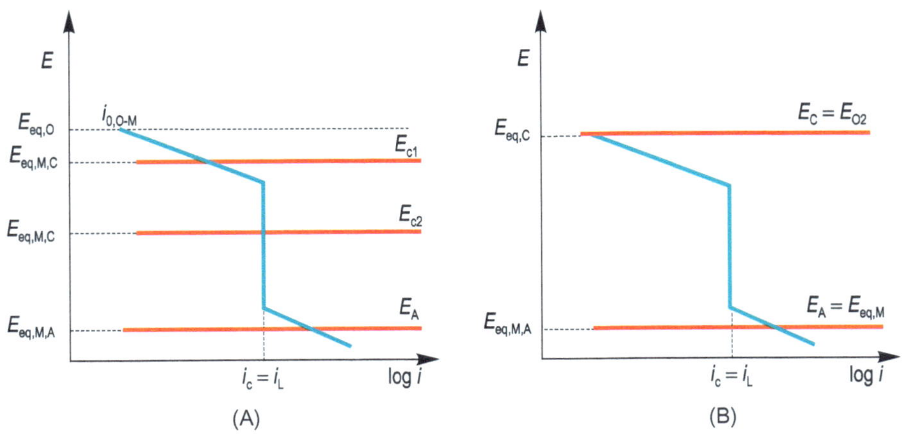

Figure 3.4 (A) Galvanic corrosion in neutral, aerated solutions for active metals and (B) galvanic corrosion in neutral, aerated solutions with noble cathode as gold.

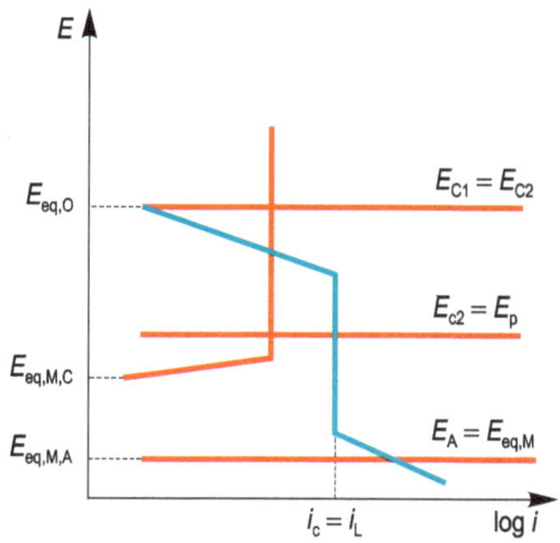

Figure 3.5 Galvanic corrosion in neutral, aerated solutions with practical noble cathode as passive metal.

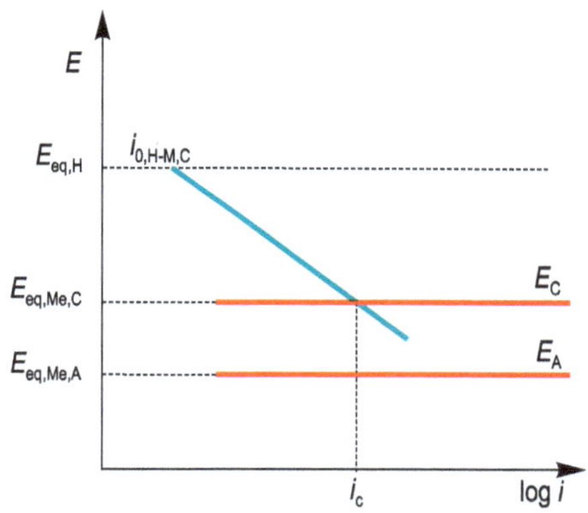

Figure 3.6 Galvanic corrosion in acids for less noble active metals.

$$E_{O_2} \cong 1.23 - 0.33 \log\left(\frac{50}{[O_2]}\right) - 0.059 \text{ pH} \tag{3.10}$$

where $[O_2]$ is the oxygen concentration in ppm.

Fig. 3.5 shows the case of passive cathodic metal. The potential of the cathode is taken as the equilibrium potential of the oxygen, given by Eq. (3.10).

3.4.1.3 Cathode potential, E_C, for hydrogen evolution

Electrochemical conditions of the galvanic coupling in acidic, oxygen-free solutions are summarized in Figs 3.6 and 3.7. When cathodic metal is less noble than hydrogen, cathode potential, E_C, is taken as equilibrium potential of the cathodic metal calculated by the Nernst equation(3.9):

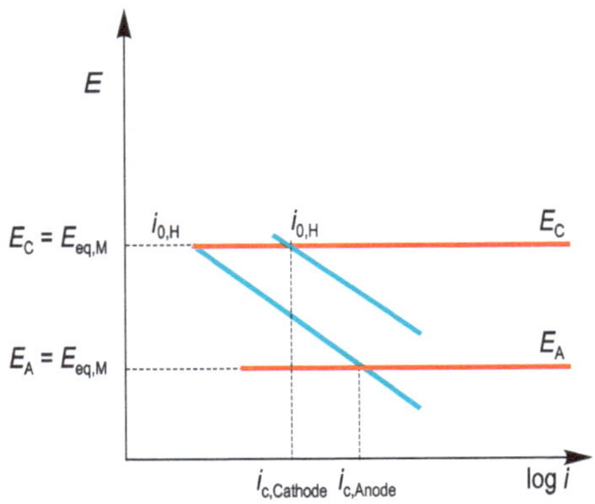

Figure 3.7 Galvanic corrosion in acids with noble active cathode.

$$E_C \cong E_{eq,M} \tag{3.11}$$

where metal ion concentration is taken as 10^{-6} mol/L (i.e. cathode does not corrode).

When cathodic metal is an active metal more noble than hydrogen (Fig. 3.7), cathode potential, E_C, is taken as equilibrium potential of the hydrogen evolution potential:

$$E_C \cong E_{H_2} = -0.059\text{pH} \tag{3.12}$$

3.4.2 Electrolyte resistivity

As rule of thumb, water resistivity can be calculated by the following equation as function of salinity and temperature as follows:

$$\rho \cong \frac{9}{(1 + 0.02\Delta T) \cdot \text{TDS}} \tag{3.13}$$

where TDS (g/L) is the total dissolved solids, and ΔT (°C) is $T - 25$.

3.4.3 Case studies for galvanic corrosion

As Figs 3.4–3.7 illustrate, following case studies can be envisaged:

- active nonnoble metals in aerated, near neutral or alkaline electrolytes (for instance Zn and Cu in seawater or soil, Fig. 3.4A, case E_{C1});
- active noble metal as cathode and an active metal as anode in aerated, near neutral or alkaline electrolytes (for instance Fe and Zn in seawater or soil, Fig. 3.4A; case E_{C2});
- noble metal as cathode and an active metal as anode in aerated, near neutral or alkaline electrolytes (for instance Au or Pt and Fe in water, Fig. 3.4B);

- passive metal as cathode and an active metal as anode in aerated, near neutral or alkaline electrolytes (for instance stainless steel or copper and Fe in seawater or soil, Fig. 3.5);
- active nonnoble metals in oxygen-free acids (for instance Zn and Fe in acids, Fig. 3.6);
- noble or passive metal as cathode and an active metal as anode in oxygen-free acids (for instance Zn and Cu in acids, Fig. 3.7).

For these galvanic couplings, the effective driving voltage and the cathodic current density are different as discussed in detail in the following sections.

3.4.3.1 Active nonnoble metals in aerated, near neutral or alkaline electrolytes (Fig. 3.4A)

Free corrosion potential are typically known; hence, driving voltage is simply their difference:

$$\Delta V = E_C - E_A \cong E_{corr,C} - E_{corr,A} \tag{3.14}$$

where symbols are known. If free corrosion potentials are not known or for a more conservative approach, the difference of the equilibrium potential can be used, by introducing proper metal ion concentration as follows:

$$E_{corr,C} \cong E_{eq,C} = E^0 + \frac{0.059}{z}\log\left[10^{-6}\right] \tag{3.15}$$

$$E_{corr,A} \cong E_{eq,A} = E^0 + \frac{0.059}{z}\log\left[10^{-2}\right] \tag{3.16}$$

As cathodic nonnoble metal does not corrode, ion concentration is the lower limit, 10^{-6} mol/L, whereas the anodic metal, because of corroding, shows a much higher concentration, typically 10^{-2} mol/L, at least. Typical reference value of E_C is E_{C2}. It should be noted that driving voltage decreases with time, as discussed later.

Cathodic current density is oxygen limiting current density (see Sections 1.5.2.3 and 2.2.2):

$$i_L \cong 10 \cdot \{[O_2] + 0.04 \cdot [Cl_2]\} \cdot \left(1 + \sqrt{v}\right) \cdot 2^{\frac{T-25}{25}} \tag{3.17}$$

i_L is expressed in mA/m^2, $[O_2]$ and $[Cl_2]$ are oxygen and chlorine contents in mg/L ($\approx$ ppm), respectively, T is the temperature in °C, and v is the water velocity in m/s.

3.4.3.2 Active noble metal as cathode and active metal as anode in aerated, near neutral or alkaline electrolytes (Fig. 3.4B)

This case study is similar to the previous one with cathodic potential typically represented by E_{C1}. Accordingly, same Eqs (3.14)–(3.16), as per previous case study, apply. Cathodic current density is again given by Eq. (3.17).

3.4.3.3 *Noble metal as cathode and active metal as anode in aerated, near neutral or alkaline electrolytes (Fig. 3.4B)*

Cathodic metal, like gold or platinum, works at the equilibrium potential of the oxygen if no other more noble cathodic process is present, whereas anode potential is equilibrium potential of the anodic active metal at a high ion concentration, typically 10^{-2} mol/L. Accordingly, Eqs (3.10), (3.15) and (3.16) apply for calculating the driving voltage and Eq. (3.17) for the cathodic current density.

3.4.3.4 *Passive metal as cathode and active metal as anode in aerated, near neutral or alkaline electrolytes (Fig. 3.5)*

Working conditions of this case study are less defined, as cathodic potential can range within the large passive interval. As in passive condition potential changes sensibly even with small current density variations, once a cathodic polarization sets up, potential drops easily to the lower limit, which is the passive potential, E_p. Two scenarios are envisaged: An initial one by which cathodic potential, E_{C1}, is equilibrium potential of oxygen reduction, given by Eq. (3.10) as shown in Fig. 3.5, and a second one to which the system evolves where cathodic potential, E_{C2}, practically coincides with passive potential or Flade potential, E_p, given by Eq. (1.23). Anode potential is equilibrium potential at a high ion concentration, typically 10^{-2} mol/L. Cathodic current density is again given by Eq. (3.17).

3.4.3.5 *Active nonnoble metals in oxygen-free acids (Fig. 3.6)*

This case study is equivalent of the one of Fig. 3.4A, in oxygen free condition. Cathodic reaction is hydrogen evolution. Driving voltage is given by same Eqs (3.14)–(3.16).

Cathodic current density, i_C, is hydrogen evolution given by means of Tafel equation as follows:

$$i_C \cong i_{0,H} \cdot 10^{\frac{-0.059\text{pH}-\left[E^0+\frac{0.059}{z}\cdot\left(\frac{T+273}{298}\right)\cdot\log a_{M^{z+}}\right]}{0.12}} \tag{3.18}$$

where $i_{0,H}$ is the exchange current density for hydrogen evolution (see Table 1.5); E^0 is the standard potential of the pertaining metal or alloy; metal ion concentration is 10^{-6} mol/L as cathode.

3.4.3.6 *Noble or passive metal as cathode and active metal as anode in oxygen-free acids (Fig. 3.7)*

Cathodic reaction is hydrogen evolution, which takes place on both metals at quite different rates: On noble metal, because working near its equilibrium potential, the rate is exchange current density of hydrogen evolution, $i_{0,H}$, of the pertaining metal (see Table 1.5) and varies strongly with metal nature.

As far as driving voltage is concerned, same Eqs (3.14)−(3.16) apply. The cathodic current density is $i_{0,H}$ of cathodic metal. With reference to Fig. 3.7, assuming surface area involved, S_C and S_A in accordance with Eq. (3.6) resulting corrosion rate is the sum of ($i_{0,cath} \cdot S_C/S_A$) and $i_{0,anode}$: In practice, acceleration of corrosion is modest because surface area ratio is low (due to the low driving voltage) and exchange current density of hydrogen evolution of cathodic metal is much lower than corrosion rate of freely corroding less noble metal.

3.4.4 Time dependency

By analysing Eq. (3.8), it appears clearly that the calculated corrosion rate is the initial maximum one, because the input driving voltage is the maximum applicable. As macrocell starts working, corrosion rate decreases with time because driving voltage lowers to reach a stationary condition. Driving voltage decreases as soon as cathode polarizes, especially when cathodic process is oxygen reduction governed by diffusion (where cathodic overvoltage relationship is a vertical line). Furthermore, when cathodic process is oxygen reduction in neutral or alkaline solutions, cathodic current density can decrease with time due to the formation of scales which reduces oxygen diffusion. Both effects contribute to reduce the galvanic corrosion rate.

This effect is well known in cathodic protection applications by galvanic anodes, typically in seawater on bare structures, where calcareous deposit forms, then reducing driving voltage as well as the oxygen limiting current density: The reduction of corrosion rate of anodes (i.e. anode consumption rate) from start-up value can exceed 90%.

3.5 Differential aeration

This type of corrosion occurring on carbon and low alloy steel, called *differential aeration*, takes place where oxygen concentration zeros, as under deposits or in cavities, because the electrolyte cannot be continuously replaced. It often takes place on buried structures (pipeline, piping, tanks) when soil is characterized by a different permeability to oxygen (Fig. 3.8): On zones where soil is less permeable to oxygen (clayey soil) corrosion occurs, whereas aerated zones (e.g. sandy soil) behave as cathode.

This mechanism occurs on carbon steel only and not on galvanized steel or copper alloys. In fact, on Cu and Zn driving voltage caused by the different oxygen concentration is small and remains constant, as Fig. 3.9 depicts. Conversely, on carbon steel, because of the alkalinity produced by the oxygen reduction on aerated zones, passivation happens so driving voltage increases, as Fig. 3.10 shows. Initial condition is ① for low oxygen zone and ② for high oxygen zone; as corrosion proceeds, zone ① starts working as anode (i.e. less noble electrode) and zone ② as cathode; hence, point ① moves to point ③ as final stable anodic corroding zone,

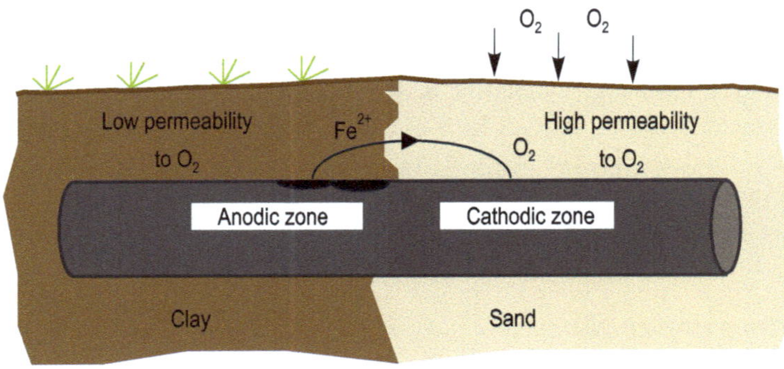

Figure 3.8 Example of differential aeration corrosion on a pipeline in soil.
From Lazzari L., Pedeferri P., Cathodic protection. Polipress. Italy: Milan; 2005.
ISBN 88-7398-020-1.

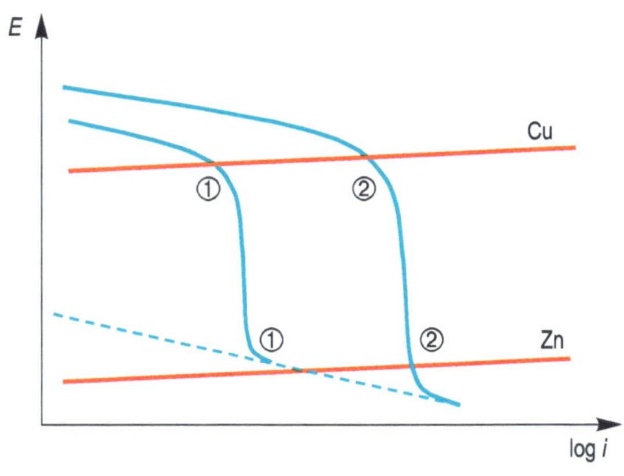

Figure 3.9 Differential aeration condition on Cu and Zn.

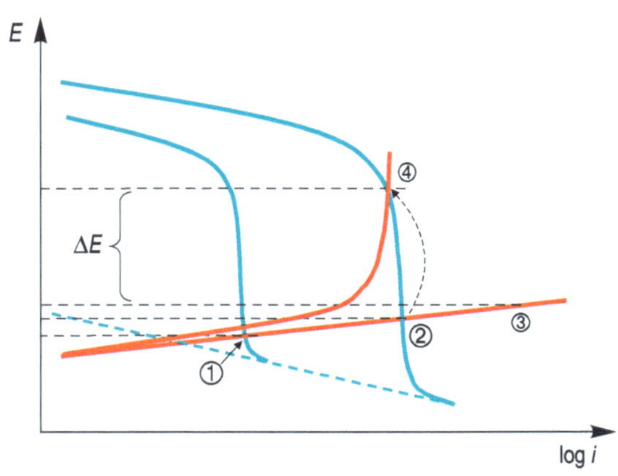

Figure 3.10 Differential aeration condition on Fe: ① low oxygen zone (initial); ② high oxygen zone (initial); ③ anodic corroding zone (final); ④ cathodic zone (final).

whereas point ② shifts to ④ because of passivation, with an increase of the driving voltage.

Corrosion rate by differential aeration, $C_{\text{rate,DA}}$, is calculated by means of the macrocell model in accordance with Eq. (3.8) which becomes, by introducing oxygen limiting current density as function of oxygen content in aerated zone and maximum expected driving voltage, ΔV, of about 0.2 V as difference between ④ and ③ of Fig. 3.10:

$$C_{\text{rate,DA}} \cong 20\sqrt{\frac{i_L \cdot \Delta V}{(\rho + 0.8)}} \cong \sqrt{\frac{[O_2]}{(\rho + 0.8)}} \tag{3.19}$$

where corrosion rate, $C_{\text{rate,DA}}$, is in mm/y; $[O_2]$ (ppm) is the oxygen content; ρ (Ωm) is the soil resistivity. Eq. (3.19) gives the initial maximum expected corrosion rate, which may decrease with time because scales form on cathodic areas, then reducing the oxygen diffusion rate.

3.6 Intergranular corrosion

This severe corrosion attack occurs on so-called sensitized alloys, typically stainless steels and nickel-based alloys. It appears as localized corrosion attack at grain boundaries, which, despite the negligible metal wastage, weakens the bond between grains, then reducing the mechanical resistance, also acting as a notch. Fig. 3.11 shows a micrograph of intergranular corrosion attack on a stainless steel, AISI 304 grade.

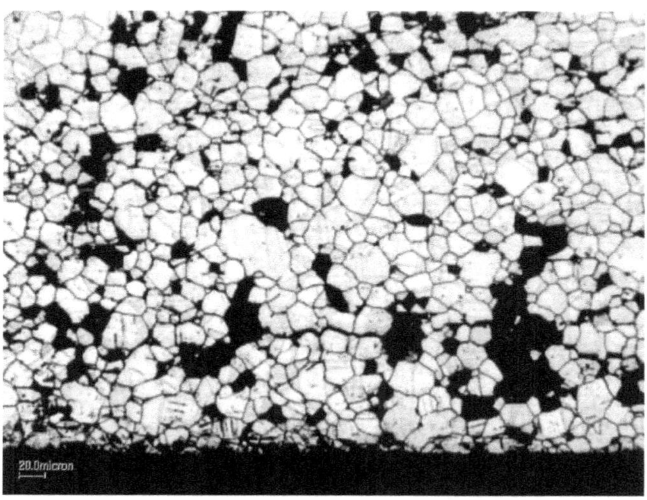

Figure 3.11 Micrograph of intergranular corrosion of AISI 304 austenitic stainless steel with 0.06% C, sensitized at 600°C for several hours.
From Pedeferri P., Corrosione e protezione dei materiali metallici. Polipress. Italy: Milan; 2007. ISBN 97888-7398-032-2 (in Italian).

Stainless steels are supplied after a *solubilization treatment* that dissolves carbon, contained at levels lower than 0.08%, in the metal matrix. This heat treatment consists of heating the alloy at 1050°C for 1 h per 1-in. thickness, to favour the dissociation of chromium carbides, then followed by a rapid cooling to avoid carbide reprecipitation. In this condition, stainless steels do not suffer intergranular corrosion.

Instead, processes such as welding can cause precipitation at grain boundaries of chromium carbides, $Cr_{23}C_6$ type, if brought for some time in a critical temperature interval (approximately 500–850°C for austenitic stainless steels; 500–700°C for ferritic stainless steels; 600–950°C for nickel alloys). This process, called *sensitization*, is the prerequisite for intergranular corrosion. Sensitization is avoided by employing *low-carbon grades*, by keeping carbon content below 0.03%, or adopting *stabilized stainless steels*.

3.6.1 Mechanism of intergranular corrosion

The precipitation of chromium carbide occurring during sensitization process (Fig. 3.12) leads to a galvanic coupling between chromium depleted zone (active behaviour) and surrounding passive zones [2]. Expected corrosion rate, $C_{\mathrm{rate,IC}}$, is given by a similar equation as (3.8):

$$C_{\mathrm{rate,IC}} \cong 20\sqrt{\frac{i_{\mathrm{C}} \cdot \Delta V}{(\rho + 0.8)}} \tag{3.20}$$

where symbols and unities are known. In low aggressive environment, corrosion rate is less than 1 mm/y, although enough to trigger mechanical failures in relatively short time.

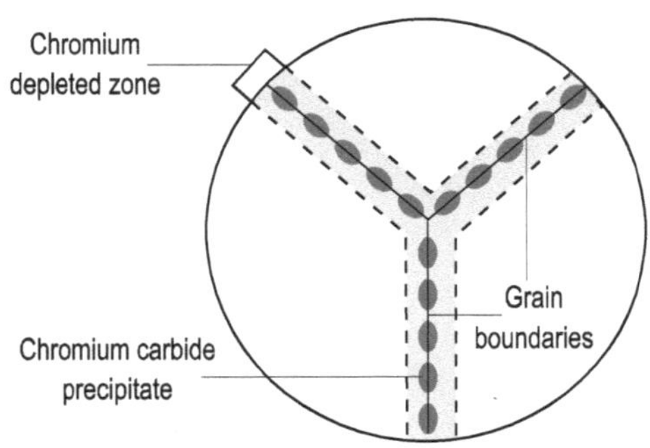

Figure 3.12 Schematic representation of chromium carbides that separate at grain boundaries.
From Pedeferri P., Corrosione e protezione dei materiali metallici. Polipress. Italy: Milan; 2007. ISBN 97888-7398-032-2 (in Italian).

3.6.2 Intergranular corrosion testing

To check sensitization of a stainless steel or a nickel alloy, laboratory tests are available, as below [3]:

- *Huey Test*, ASTM A262 – Practice C. Testing solution is a boiling nitric acid 65% solution and testing time consists of five periods, each of 48 h; corrosion rate is calculated from weight loss measurements.
- *Strauss Test*, ASTM A262 – Practice E. Testing solution is a boiling $Cu/CuSO_4$ – 16% sulphuric acid solution. After exposure for 1 h, the specimen is 180 bent over a rod with diameter equal to the specimen thickness and visually examined: no cracks allowed.
- *Streicher Test*, ASTM A262 – Practice B. Testing solution is a boiling $Fe_2(SO_4)_3$ – 50% sulphuric acid solution for 24 to 120 h. As per Huey test, corrosion rate is calculated from weight loss measurements.

In all tests listed above, a common cathodic process is hydrogen evolution; in Strauss and Streicher two more cathodic processes occur, as copper ion reduction to copper and ferric ions reduction to ferrous ions, respectively. Therefore, it appears that only in Huey test corrosion rate may be predicted.

Let's consider solubilized condition first. In Huey test, stainless steel is passive so corrosion rate coincides with passivity current density at severe conditions as established by the strong acidity (pH zero) and temperature above 100°C. Accordingly, expected corrosion rate is typically less than half mm/y.

For sensitized alloys, intergranular corrosion rate is given by Eq. (3.20), in which cathodic current density is hydrogen evolution in accordance with Eq. (3.18) by inputting: $a_{M^{z+}} = 1$ M; $\Delta V = 0.1$ V; pH = 0; $i_{0,H} = 10$ mA/m^2. It results that sensitized alloy exhibits a corrosion rate of about 30 mm/y, easily measured by weight loss even in a short duration test.

3.7 Summary

Table 3.2 summarizes equations for the calculation of corrosion rates of localized corrosion attacks given by galvanic coupling, differential aeration and intergranular corrosion.

3.8 Appendix

3.8.1 Coefficient of variation, CV

As discussed in Chapter 8, Statistical Analysis of Corrosion Data, corrosion rates calculated by an algorithm have a stochastic meaning, because in practice results of a corrosion process are spread in a distribution.

The use of the calculated value, as well as for measured values in testing or in inspections, should take into account this interpretation. As said, CV derived from

Table 3.2 Equations for calculation of localized corrosion rate for galvanic coupling, differential aeration and intergranular attack

	Equation	Parameters
Macrocell model	$C_{rate,MC} = k \cdot i_C \cdot \frac{S_C}{S_A}$	$[Cl_2]$ (ppm) chlorine content $C_{rate,DA}$ (mm/y) corrosion rate $C_{rate,GC}$ (mm/y) corrosion rate $C_{rate,IC}$ (mm/y) corrosion rate $C_{rate,MC}$ (mm/y) corrosion rate E_A (V) potential of anode E_C (V) potential of cathode E_{H_2} (V) potential of hydrogen evolution E_{O_2} (V) potential of oxygen reduction ΔV (V) driving voltage ϕ_k (m) characteristic dimension i (A/m^2) average current density i_C (A/m^2) cathodic current density i_L (mA/m^2) oxygen limiting current density $i_{0,H}$ (mA/m^2) exchange current density of hydrogen on metal k (m$^{-0.5}$) dimensional constant
Throwing power (see Table 4.1 for case studies)	$L_{max} = k\sqrt{\frac{\Delta V \phi_k}{\rho_i}}$	
Maximum Surface Area Ratio in Macrocell	$\left(\frac{S_C}{S_A}\right)_{max} \cong \left(1+\frac{S_C}{S_A}\right)_{max} \cong k\sqrt{\frac{\Delta V}{i_C \cdot (\rho + 0.8)}}$	
Galvanic corrosion rate	$C_{rate,GC} \cong 20\sqrt{\frac{i_C \cdot \Delta V}{(\rho + 0.8)}}$	
Potential of cathodic processes (V SHE)	$E_{O_2} \cong 1.23 - 0.059[pH + (10 - [O_2])]$ $E_{H_2} = -0.059pH$	
Resistivity of fresh waters (Ωm)	$\rho \cong \frac{9}{(1+0.02\Delta T) \cdot TDS}$	
Galvanic coupling in aerated solutions, active metals	$\Delta V = E_C - E_A \cong E_{corr,C} - E_{corr,A}$ $E_{corr,C} \cong E_{eq,C} = E^0 + \frac{0.059}{z}\log[10^{-6}]$ $E_{corr,A} \cong E_{eq,A} = E^0 + \frac{0.059}{z}\log[10^{-2}]$ $i_C = i_L \cong 10 \cdot \{[O_2] + 0.04 \cdot [Cl_2]\} \cdot (1+\sqrt{v}) \cdot 2^{\frac{T-25}{25}}$	
Galvanic coupling in aerated solutions, with a passive metal	$\Delta V = E_C - E_A \cong E_p - E_{eq,M}$ $E_p \cong \left(\frac{298}{T+273}\right) \cdot \left[0.63 + 0.06\left(\frac{pH-7}{5}\right) - \frac{PREN}{100} - 0.05\log\left(\frac{Cl}{36}+1\right)\right]$ $E_A \cong E_{eq,M} = E^0 + \frac{0.059}{z}\log[10^{-2}]$ $i_C = i_L 0$	

Galvanic coupling in oxygen-free acids, with active metals	$\Delta V = E_C - E_A \cong E_{corr,C} - E_{corr,A}$ $E_{corr,C} \cong E_{eq,C} = E^0 + \frac{0.059}{z}\log[10^{-6}]$ $E_{corr,A} \cong E_{eq,A} = E^0 + \frac{0.059}{z}\log[10^{-2}]$ $i_C \cong i_{0,H} \cdot 10^{\frac{-0.059pH-\left[E^0+\frac{0.059}{z}\cdot\left(\frac{T+273}{298}\right)\cdot\log a_{M^{z+}}\right]}{0.12}}$	L_{max} (m) throwing power $[O_2]$ (ppm) oxygen content ρ (Ωm) electrolyte resistivity PREN = %Cr + 3.3% Mo + X%N X = 16 for austenitic stainless steels X = 30 for duplex stainless steels S_C (m^2) cathodic surface area S_A (m^2) anodic surface area TDS (g/L) salinity ΔT (°C) = T-25 T (°C) temperature v (m/s) water velocity
Galvanic coupling in oxygen-free acids, with a passive metal	$\Delta V = E_C - E_A \cong E_{corr,C} - E_{corr,A}$ $E_{corr,C} \cong E_{eq,C} = E^0 + \frac{0.059}{z}\log[10^{-6}]$ $E_{corr,A} \cong E_{eq,A} = E^0 + \frac{0.059}{z}\log[10^{-2}]$ $i_C = i_{0,H}$	
Differential aeration (mm/y)	$C_{rate,DA} \cong 20\sqrt{\frac{i_L \cdot \Delta V}{(\rho+0.8)}} \cong \sqrt{\frac{[O_2]}{(\rho+0.8)}}$	
Intergranular corrosion of stainless steel (mm/y)	$C_{rate,IC} \cong 20\sqrt{\frac{i_C \cdot \Delta V}{(\rho+0.8)}}$	

the experience can be suitably used for further calculations such as the maximum expected value.

When dealing with localized corrosion, in which macrocell mechanism applies, the following CV values can be adopted:

- *Galvanic corrosion*: CV = 0.2 (in some instance, 0.4 could also be considered). The maximum expected value can be taken as $\mu + 2\sigma$, where the standard deviation, σ, is obtained from the coefficient of variation, CV. By assuming a CV equals to 0.2, the maximum value is $x_{max} \cong 1.4\ \mu$;
- *Differential aeration corrosion*: CV = 0.4 (in some instance, 0.6 could also be considered). The maximum expected value can be taken as $\mu + 2\sigma$, where the standard deviation, σ, is obtained from the coefficient of variation, CV. By assuming a CV equals to 0.4, the maximum value is $x_{max} \cong 1.8\ \mu$. On the basis of Gumbel statistics, $x_{max} = 2.25\ \mu$;
- *Intergranular corrosion*: CV = 0.2. By adopting extreme values statistics, maximum value by Gumbel statistics is $x_{max} = \mu \cdot (1 + 3.14\ \text{CV})$; therefore $x_{max} = 1.63\ \mu$.

3.8.2 Throwing power in CP by galvanic anodes

In seawater, CP by galvanic anodes is an example of galvanic corrosion in which protection condition of the cathode is of interest rather than the wastage of the anode [1]. The design of CP for typical applications as platforms and piers is based on the calculation of the throwing power or the equivalent surface area ratio, S_C/S_A, which practically and luckily does not change with time, although affecting parameters vary strongly. By applying Eq. (3.6), the initial S_C/S_A ratio is about 50 because driving voltage at the beginning approximates half a volt and oxygen limiting current density exceeds 100 mA/m^2. Once calcareous deposit has formed, as consequence of alkalinization of the cathode, driving voltage drops to about one-tenth of the initial value and the same for oxygen limiting current density which also drops to a tenth, so S_C/S_A ratio remains practically unchanged. This condition helped achieve an acceptable empirical design much before the most correct and sophisticated design based on modelling (see Chapter 6: Computer Modelling).

3.8.3 Galvanic corrosion vs differential aeration in soil

On buried structures, when severe corrosion attacks occur there are three possible causes:

- interference or stray current corrosion;
- galvanic corrosion;
- differential aeration corrosion.

The first one is difficult to predict since effecting conditions are often undefined [1]. Conversely, as discussed above, galvanic and differential aeration corrosion rates are predictable. In short, the same general equation applies, but result is different because driving voltage is different. In differential aeration, maximum driving voltage is 200 mV, often around 100 mV, whereas in galvanic coupling, it can exceed also a half volt. Accordingly, with the same other conditions, i.e. oxygen concentration and resistivity, galvanic corrosion rate is at least twice the one by differential aeration.

3.8.4 Galvanic corrosion risk in clad pipelines

In oil-&-gas activities, when expected corrosion conditions are severe, it is common practice to use internally clad pipes, with cladding materials a Ni-based alloy, typically Alloy 625 (Ni 58% min; Cr 20%–23%; Fe 5%; Mo 8%–10%; Nb + Ta 3.15%–4.15%; PREN 51). Ni-based alloy resists all expected corrosion attacks: uniform corrosion; pitting corrosion; sulphide stress corrosion cracking and hydrogen-induced cracking. Nevertheless, the following two typical case studies are frequent.

3.8.4.1 Presence of a defect in the overlay

In the presence of a defect on the overlay or for local chromium depletion (typically after welding repair), galvanic corrosion condition sets up between backing carbon steel and surrounding passive Ni alloy in many exposure conditions from hydrotesting to operating.

Galvanic corrosion rate is estimated as follows:

- maximum surface area ratio S_C/S_A is given by Eq. (3.6). Assuming a roughly constant electrolyte resistivity for various exposure conditions, the ratio is strongly influenced by the cathodic process (oxygen reduction in aerated conditions or acidic oxygen-free conditions);
- galvanic corrosion rate is given by Eq. (3.8).

For example, in stagnant aerated seawater (resistivity 0.25 Ωm; oxygen limiting current density 0.1 A/m^2; driving voltage 1 V as maximum), corrosion rate is 6 mm/y, when assuming prudentially a driving voltage of 0.5 V.

In the presence of oxygen-free formation water, same resistivity as seawater, in equilibrium with a gas containing CO_2 with partial pressure of 5 bar, which can lower pH to about 3.6, cathodic reaction is hydrogen evolution, whose maximum cathodic current density is exchange current density of hydrogen evolution on Ni, which is 0.01 A/m^2. Driving voltage is 0.22 V as obtained from Fig. 3.7 ($E_C = -0.22$ V SHE; $E_A = -0.44$ V SHE, iron ion concentration close to molarity): This is in accordance with Battelle Report [4] which indicates a maximum value of 0.2 V. Hence, expected galvanic corrosion rate is close to 1 mm/y. On the anode, acidic corrosion, due to the presence of carbon dioxide, sets up at a corrosion rate of about 2.6 mm/y, as reported in Table 2.A1. In this case study, galvanic effect is much lower than acidic corrosion. As a conclusion, this result confirms that galvanic coupling in acidic, oxygen-free electrolytes causes a corrosion rate about an order of magnitude lower than in aerated solutions.

3.8.4.2 Temporary coupling with carbon steel

Another case study is the galvanic coupling, often temporary, between a clad pipe and a bare carbon steel pipe, end-to-end welded, full of stagnant hydro-testing water, either fresh or sea water. Cathodic process is oxygen reduction in stagnant condition; hence, oxygen limiting current density can be easily estimated through Eq. (3.17). Two main questions arise: What would be the surface area ratio and the extension of the anodic zone for the coupling, taking into account that a general corrosion affects the carbon steel due to the presence of some oxygen. It appears evident that the maximum surface area ratio in accordance with Eq. (3.6) does not apply, as both cathodic and anodic areas can extend on each side until total available driving voltage is consumed. As predicted by the model proposed, the only dissipation is from the cathodic side and throwing power can be estimated by the equation reported in Table 3.1, for "inside-a-tube" geometry. By this, cathodic surface area is determined. As overvoltage on the anodic, corroding zone is taken negligible, the relevant coupled anodic zone is not less the cathodic one; hence, as a maximum, the surface area ratio is unitary. Therefore, on carbon steel section, adjacent the clad section, corrosion rate doubles, as a maximum, for a length of the same order of the throwing power; from an engineering viewpoint, it can be considered as a negligible effect, provided that oxygen content is controlled, most likely below 1 ppm.

3.8.5 Intergranular corrosion rate prediction

3.8.5.1 Solubilized nonsensitized alloy

In acid solution, passive metals can suffer uniform corrosion if depassivation takes place. For stainless steels and nickel-based alloys, this occurrence can happen when pH is below 2.

Without depassivation, two conditions are possible:

1. No chlorides. Passivity current density of many stainless steels and nickel alloys is less than exchange current density of hydrogen evolution; therefore, free corrosion potential is equilibrium potential of hydrogen evolution ($-0.059\,\text{pH}$ V SHE) and passivity current density is lesser than $i_{0,\text{H}}$;
2. Presence of chlorides. Passivity current density in the presence of chlorides is in the same range of exchange current density of hydrogen evolution, free corrosion potential is as a maximum about 100 mV more negative than hydrogen equilibrium potential (Tafel slope is 0.12 V/decade) as depicted in Fig. 3.A1. Passivity current density can be estimated from the cathodic current density of hydrogen evolution, which is easily calculated by:

$$i_{\text{H}} \cong i_{0,H} \cdot 10^{\frac{-0.059\text{pH}-E_{\text{corr}}}{0.12}}$$

where $i_{0,\text{H}}$ is the exchange current density for hydrogen evolution (in the range $1-10\ \text{mA/m}^2$), and E_{corr} is the free corrosion potential of passive metal in acidic solution.

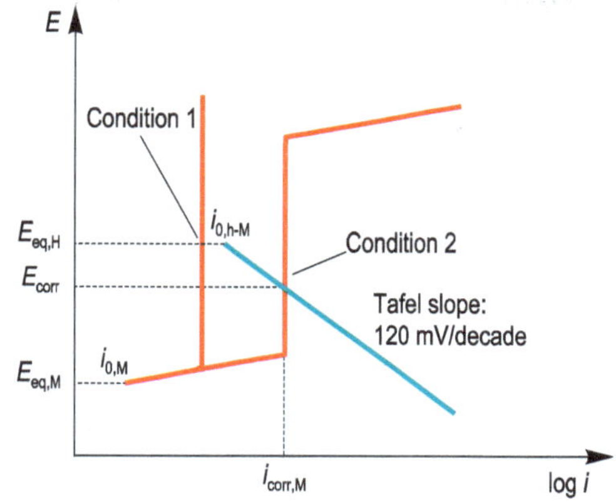

Figure 3.A1 A simplified model for acid corrosion of passive metals (with and without chlorides).

3.8.5.2 Sensitized alloy

Intergranular corrosion rate of sensitized alloys in Huey test is given by Eq. (3.20), in which cathodic current density is hydrogen evolution in accordance with Eq. (3.18). Inputs are as follows: $\Delta V = 0.1$ V, pH = 0, $\rho = 1\ \Omega$m, $i_{0,\mathrm{H}} = 10$ mA/m^2. It results that sensitized alloy exhibits in Huey test a corrosion rate of about 30 mm/y, easily measured by weight loss.

When exposed to the process fluid, which shows different corrosion conditions, intergranular corrosion rate of sensitized alloy is lower although not negligible, given by the same Eq. (3.20). For instance, pH = 3.5, $T = 85°$C, electrolyte resistivity, $\rho = 1\ \Omega$m; other input data are as follows:

driving voltage, ΔV, given by:

$$\Delta V = E_\mathrm{C} - E_\mathrm{A} \cong -0.059\ \mathrm{pH} - \left[E^0 + \frac{0.059}{z}\left(\frac{T+273}{298}\right)\cdot \log a_{\mathrm{M}^{z+}}\right]$$

where $a_{\mathrm{M}^{z+}} = 2\ \mathrm{mol}/L$ inside intergranular attack;

cathodic current density, i_H, by Tafel Law is conservatively calculated at anode potential, therefore:

$$i_\mathrm{H} \cong i_{0,H} \cdot 10^{\frac{-0.059\mathrm{pH}-E_\mathrm{A}}{0.12}}$$

Rounding the driving voltage to 50 mV, cathodic current density is 26 mA/m^2, giving an intergranular corrosion rate of about 0.7 mm/y, sufficient to initiate pitting or trigger SCC when conditions for occurrence exist.

References

[1] Lazzari L, Pedeferri P. Cathodic protection. Milan, Italy: Polipress; 2005. ISBN 88-7398-020-1.
[2] Pedeferri P. Corrosione e protezione dei materiali metallici. Milan, Italy: Polipress; 2007. ISBN 97888-7398-032-2 (in Italian).
[3] ASTM A262-02. Standard practices for detecting susceptibility to intergranular attack in austenitic stainless steels. PA: ASTM International; 2002.
[4] Battelle Report. Stainless and nickel-base alloys for sour reservoirs – volume I: experimental results. TX: Final Report, Battelle Petroleum Technology Centre; 1985.

Pitting and Crevice Corrosion

4

These forms of corrosion proceed by two distinguished stages (Fig. 4.1):

- initiation (or induction time);
- propagation.

Propagation occurs through a macrocell mechanism, where cathodic and anodic processes are separated, respectively on cathodic and anodic areas, characterized by a high surface area ratio ($S_C \gg S_A$). Once pitting or crevice attack has started, the propagation cannot be halted unless cathodic process is totally removed or by applying cathodic protection.

4.1 Initiation stage for pitting

The initiation stage represents the time required for the local breakdown of passive film which is produced by the action of specific chemical species, present in the environment, such as chloride ions (Cl^-) and in less extent halides (Br^-, F^-, I^-). It is agreed that the necessary electrochemical condition required to breakdown locally the passive film is that potential of cathodic process must be more noble than a specific operational parameter, named *pitting potential*, which depends on metal and environmental properties. This step might last a few weeks up to several months, depending on metal composition and operating conditions; as rule of thumb, initiation time, also called incubation time, never exceeds a year.

In fact, pitting can be categorized as an infant mortality phenomenon because should it start due to favourable conditions, the local breakdown of passive film takes place in early time after exposure. Conversely, if a critical period of time has elapsed, pitting never starts anymore, unless operating conditions worsen: should this happen, induction time starts again based on newer conditions.

Typically, pit starts where passive film is weaker or flawed, for example, near welding because of depletion of some elements or because oxide film is too thin or because superficial inclusions jeopardize the integrity of passive film. Surface finishing strongly influences pitting initiation: smooth surfaces are more resistant or result into few, large pits, whereas rough surfaces experience easier pitting initiation of numerous small pits. Stagnant condition favours pitting initiation, whereas agitation or turbulence condition helps inhibit it.

Another important aspect related to pit initiation is the presence of biofilm as recognized since the 1960s: Sterile electrolytes do not cause pitting, whereas natural seawater does, because bacterial activity raises the potential of the cathodic process above pitting potential of many stainless steels.

Fig. 4.2A and B shows examples of pitting attack on stainless steel.

Engineering Tools for Corrosion. DOI: http://dx.doi.org/10.1016/B978-0-08-102424-9.00004-5

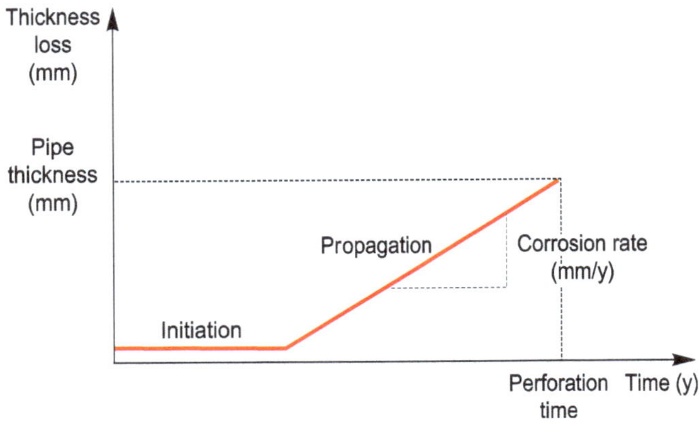

Figure 4.1 Stages of initiation and propagation of localized corrosion.
From Pedeferri P., Corrosione e protezione dei materiali metallici. Polipress. Italy: Milan; 2007. ISBN 97888-7398-032-2 (in Italian) [1].

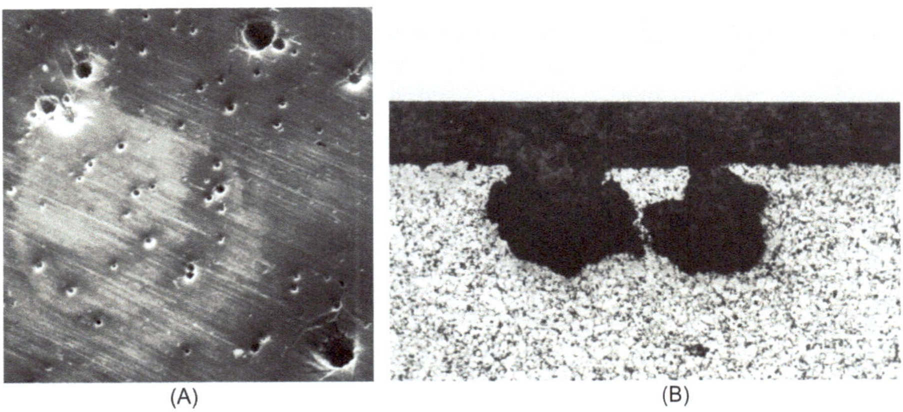

Figure 4.2 (A) Pitting attacks on AISI 304 stainless steel. (B) Typical cavern shape of pitting attack on AISI 304 stainless steel.
(A) and (B) From Pedeferri P., *Corrosione e protezione dei materiali metallici*. Polipress. Italy: Milan; 2007. ISBN 97888-7398-032-2 (in Italian).

4.1.1 Electrochemical condition for pitting initiation

Passive film breakdown occurs when driving voltage, ΔE, as difference between the potential of cathodic reaction occurring on passive film and pitting potential ($\Delta E = E_C - E_{pit}$) is positive. This condition is represented on zone A in Pedeferri's Diagrams as depicted in Fig. 4.3. In early 1990s, Pedeferri proposed a potential-chloride content plot, $E - \log[Cl]$, to predict corrosion condition for steel reinforcement in concrete [2] and postulated that a similar diagram applies to stainless steels,

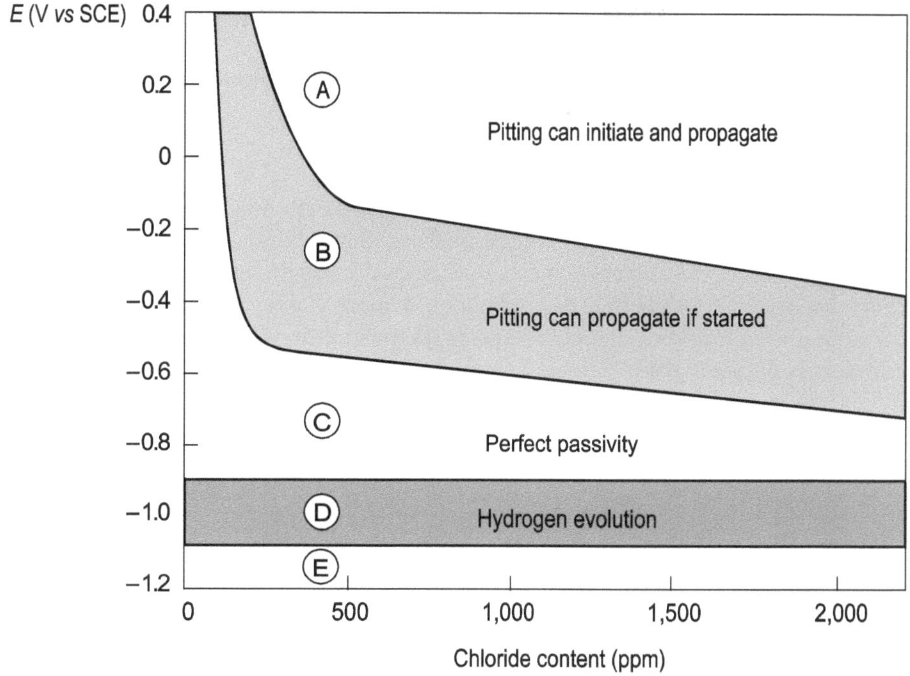

Figure 4.3 Pedeferri's diagram for passive carbon steel and stainless steel [2].

as the research group *PoliLaPP* (Corrosion Laboratory "*Pietro Pedeferri*" at Politecnico di Milano, Milan, Italy) is developing.

4.1.1.1 Potential of the cathodic process, E_C

Potential of cathodic reaction, E_C, is given by the following three conditions occurring in most industrial-related environments:

- oxygen reduction in sterile electrolyte;
- oxygen reduction in the presence of biofilm;
- chlorine reduction.

Potential of oxygen reduction in sterile electrolyte is the potential obtained by the Nernst equation, therefore function of pH and oxygen concentration according to Eq. (1.21):

$$E_{O_2} \cong 1.23 - 0.33 \log\left(\frac{50}{[O_2]}\right) - 0.059 \text{ pH} \tag{4.1}$$

where O_2 is the oxygen concentration in ppm.

In the presence of biofilm, such as in seawater, potential of the oxygen reduction reaction can be expressed simply by adding +0.2 to +0.3 V to the potential in absent of biofilm. This ennoblement of the potential in living seawater was recognized at an early stage by Defranoux [3] and later measured and attributed to the presence of a biofilm by Mollica et al. [4]. Therefore, as minimum:

$$E_{O_2/\text{bio}} = E_{O_2} + 0.2 \tag{4.2}$$

This potential ennoblement is the principle on which so-called biofilm probes are based for monitoring biofilm formation in seawater circuits.

In the presence of chlorine, such as in treated or sanitized waters, potential of the cathodic reaction of chlorine reduction is more noble than oxygen reduction therefore pitting initiation can occur also in absence of oxygen. Potential of chlorine reduction is obtained from Nernst equation as follows:

$$E_{Cl_2} = 1.36 + 0.6 \log\left[\frac{Cl_2}{36}\right] \tag{4.3}$$

where Cl_2 is the chlorine concentration in ppm (>0.1).

4.1.1.2 *Pitting potential,* E_{pit}

Pitting potential, E_{pit}, is function of temperature, chloride concentration, pH and stainless steel composition. From experience confirmed by laboratory tests, the following parametric equation is proposed as addition of logarithm of influencing parameters:

$$E_{\text{pit}} \cong \frac{298}{T+273} \cdot \left(\frac{\text{PIN}}{25} + \log(1+v) - 0.25 \cdot \log\left[\frac{\text{Cl}}{36} + 1\right] + \frac{\text{pH}-7}{25}\right) \tag{4.4}$$

where E_{pit} is the pitting potential in V SHE; T (°C) is the temperature; v (m/s) is the fluid velocity; [Cl] (ppm) is the chloride concentration. Coefficients of chloride and pH terms are based on the molar ratio of 0.6 that influences passivity condition. PIN, *Pitting Initiation Number*, is function of stainless steel composition:

$$\text{PIN} = \text{PREN} - 0.1\%\text{Mn} - 100\%\text{S} \tag{4.5}$$

$$\text{PREN} = \%\text{Cr} + 3.3\%\text{Mo} + k\%\text{N} \tag{4.6}$$

Coefficient of nitrogen, k, is 16 or 30 or 0 for austenitic or duplex or ferritic stainless steels, respectively.

PIN differs slightly from PREN. In fact, it involves Mn and S content that gives an indirect indication of the amount of sulphide inclusions, which triggers film breakdown when present at the metal surface.

4.1.2 Pitting equation

Eq. (4.4) can be rewritten in a state equation form as follows:

$$E_{pit}\cdot (T+273) \cong 298 \log\left(2.5\cdot \prod_i Y_i\right) \quad (4.7)$$

where E_{pit} is the pitting potential (V SHE), T (°C) is the temperature, and Y_i is the parameter relating to affecting factors, namely PIN, flow rate, chloride content and pH, i.e., stainless steel composition and operating conditions. Constant 2.5 is the for SHE scale. Eq. (4.7) can be named *pitting equation.*

4.1.3 Pitting induction time (PIT) and general pitting equation

From experience, time required for pitting initiation, which is in turn time required to breakdown locally the passive film, once established the electrochemical condition $E_C > E_{pit}$, is given by an exponential equation as follows:

$$\mathrm{PIT} \cong k\cdot 10^{\left(\frac{\mathrm{PREN}}{2\log[\mathrm{Cl}]}(1-\Delta E_{pit})\right)} \quad (4.8)$$

where PIT (h) is the time; k (h) is an experimental constant generally close to 1; [Cl] (ppm) is the chloride concentration (>10 ppm); ΔE_{pit} (V) is driving voltage ($E_C - E_{pit}$; where E_C is the potential of the cathodic process and E_{pit} is pitting potential calculated by Eq. (4.4)). As said, as pitting is an infant mortality phenomenon, in practice when PIT exceeds 10^4 h pitting does not initiate.

Eq. (4.8), which can be named *general pitting equation*, offers a comprehensive relationship between stainless steel properties (PREN or PIN), electrolyte operating conditions and induction time. In details, general pitting equation establishes a relationship between:

- PIT (pitting induction time);
- E_{pit}, pitting potential or potential breakdown of passive film (it depends on the energy barrier for passive film disruption);
- E_c, potential of cathodic process, which can be in general:
  - E_{O_2}, for oxygen reduction reaction;
  - $E_{O_2/bio}$, for oxygen reduction reaction in the presence of biofilm;
  - E_{Cl_2}, for chlorine reduction reaction;
- PIN (which represents, like PREN, the strength or robustness of passive film);
- [Cl] chloride concentration (chloride ions are in competition with hydroxyls: the former weaken the film, the latter strengthen it. To establish the prevailing one, the molar ratio [Cl]/[OH] is generally considered: larger than 0.6 passive film breaks, smaller the passive film resists);
- pH (acidic conditions weaken the passive film, conversely alkaline conditions strengthen it);
- T, temperature (the higher the temperature the lower the passive film resistance);
- v, fluid velocity (or stagnant condition) in aerated chloride containing solutions; high fluid velocity favours stable passivity.

4.1.4 Critical-chloride-concentration (for pitting and crevice)

From laboratory testing results, *critical-chloride-concentration* (CCC) for either pitting or crevice is a function of stainless steel composition, i.e. PIN, pH and temperature, as follows:

$$\log[\mathrm{Cl}^-]_{\mathrm{critical}} \cong \left[\frac{\mathrm{PIN}}{9.5} + \frac{\mathrm{pH}-7}{5} - \frac{T-25}{100}\right] \quad (4.9)$$

where parameters are known.

4.2 Initiation stage for crevice

In crevice corrosion of stainless steels, initiation stage follows two steps:

1. oxygen depletion in the crevice;
2. passivity breakdown inside crevice.

First step takes time and is called oxygen depletion stage or crevice-incubation time. Primarily, it depends on crevice gap width: If higher than a critical threshold, named *crevice critical gap size* (CCGS), electrolyte inside crevice is renewed so no oxygen depletion occurs. From literature data [5,6], CCGS for stainless steels, assuming the crevice 5 mm deep, can be estimated by the following equation based on stainless steel composition through PREN:

$$\mathrm{CCGS} \cong \left(\frac{17}{\mathrm{PREN}}\right)^2 \quad (4.10)$$

where CCGS is in μm and ranges between 0.1 μm and 0.1 mm. As rule of thumb, the narrower and the deeper the gap (i.e. the more closed geometry) the more likely the crevice occurrence.

Inside the gap, oxygen is consumed through the corrosion reactions occurring on passive stainless steel: oxygen reduction and passive film growth as follows:

$$O_2 + 2H_2O + 4e = 4OH^- \quad (4.11)$$

$$x\mathrm{M} + yH_2O = M_xO_y + 2yH^+ + 2y\,e \quad (4.12)$$

the rate of reactions coincides with the slowest one which is the passivity current density of the stainless steel, i_p (mA/m^2), which is a function of PREN, pH, temperature and chlorides (refer to Chapter 1: Basic Principles, Eq. (1.25)). Accordingly, the *Crevice Induction Time*, *CIT*, is proportional to the inverse of passivity current, hence function of same influencing parameters (i.e. PREN, pH, temperature and chlorides). During this stage inside crevice, whereas oxygen depletes, pH decreases as per

Eq. (4.11) and potential of metal inside crevice lowers, hence triggering a macrocell whose anode is inside crevice and the cathode is the passive external surface.

The macrocell current causes an increase of the potential inside crevice leading to two possible consequences:

- passive film strengthen, so crevice attack does not start, as it occurs for high PREN stainless steels;
- passive film breakdowns because the breakdown potential of passive film, E_{rup}, is reached so crevice attack starts propagating following the same mechanism of pitting.

Breakdown potential of passive film, E_{rup}, is similarly to pitting potential, the potential at which the passive film inside crevice breaks. As inside crevice pH lowers to a value close to two and there is no effect of velocity of the electrolyte, breakdown potential of a passive film, E_{rup}, inside crevice is given by the following equation, derived from pitting potential Eq. (4.4):

$$E_{rup} \cong \frac{298}{T+273} \cdot \left(\frac{PREN}{25} - 0.25 \cdot \log\left[\frac{Cl}{36}+1\right] - 0.20 \right) \quad (4.13)$$

where symbols are known.

By this interpretation, crevice starts because of oxygen depletion first and then because of the presence of a driving voltage which sets up the macrocell; Oldfield gave another interpretation based on passive film breakdown as only due to the pH decreasing inside crevice below a critical value [5,6].

The stage of crevice initiation depends on all mentioned parameters, namely: PREN, pH, temperature, chlorides content and driving voltage as difference between potential of cathode process outside crevice and breakdown potential of passive film. The relationship which includes all parameters is the following:

$$CIT \cong k \cdot 10^{\frac{PREN}{3\log[Cl]}(1-\Delta E_C)} \quad (4.14)$$

where CIT (h) is the crevice induction time; k (h) is an experimental constant close to 1; [Cl] (ppm) is the chloride concentration (>10 ppm); ΔE_C (V) is the driving voltage ($E_C - E_{rup}$; where E_C is potential of the cathodic process and E_{rup} is calculated by Eq. (4.13)). As rule of thumb, when CIT exceeds 10^4 h or approximately a year, crevice does not start.

In summary, conditions for crevice occurrence are the following:

- chloride concentration higher than critical threshold (necessary but not sufficient);
- gap width smaller than CCGS (necessary but not sufficient);
- CIT shorter than a year (the shorter CIT the higher the likelihood for crevice occurrence). As rule of thumb, when calculated incubation time exceeds 1 year, likelihood of crevice initiation zeros.

These conditions are in AND relation, therefore crevice is prevented by avoiding just one. A further and definitive prevention method is the application of CP which prevents or inhibits the setup of the macrocell.

4.3 Propagation

As macrocell current becomes stationary, metal dissolves inside pit or crevice and surrounding passive zones work as cathode. Inside either a pit or a crevice, the solution becomes gradually more aggressive as hydrolysis reaction of metal ions proceeds so acidification increases and pH drops to values less than 3 in pitting and less than 2 in crevice:

$$M^{z+} + zH_2O \rightarrow M(OH)_z + zH^+ \tag{4.15}$$

Because of this environmental change, ripassivation inside a pit and a crevice is not possible; conversely, on cathodic zones, pH increases, then passive film strengthens.

Macrocell current is regulated by the field equation as discussed in Chapter 3, Localized Corrosion and throwing power is the governing parameter. For plane geometry, in aerated solutions, the maximum corrosion rate is given by the following general equation:

$$C_{\text{rate,MC}} \cong 20\sqrt{\frac{i_L \cdot \Delta V}{(\rho + 0.8)}} \tag{4.16}$$

where $C_{\text{rate,MC}}$ (mm/y) is the corrosion rate, ΔV (V) is the driving voltage, ρ (Ω m) is the electrolyte resistivity, i_L (A/m^2) is the oxygen limiting current density.

For pitting on stainless steels, driving voltage ΔV ranges between 0.5 and 1 V, therefore, maximum expected corrosion rate is derived from (4.16):

$$C_{\text{rate,pitting}} \cong 20\sqrt{\frac{i_L}{(\rho + 0.8)}} \tag{4.17}$$

For crevice on stainless steels, driving voltage ΔV ranges between 0.1 and 0.25 V, because of the high ohmic drop across the crevice. Therefore, for maximum expected crevice corrosion rate Eq. (4.16) shorts to

$$C_{\text{rate,crevice}} \cong 10\sqrt{\frac{i_L}{(\rho + 0.8)}} \tag{4.18}$$

Figs. 4.4 and 4.5 depict the macrocell for pitting and crevice attack propagation.

4.4 Summary

Table 4.1 summarizes equations proposed for the calculation of initiation time and propagation rate for pitting and crevice.

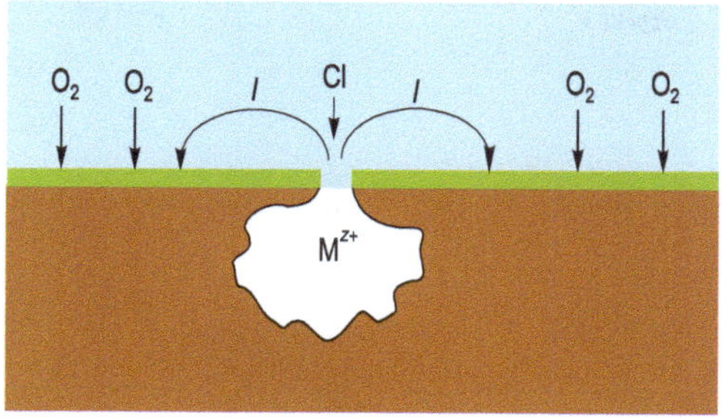

Figure 4.4 Scheme of mechanism for pitting propagation.

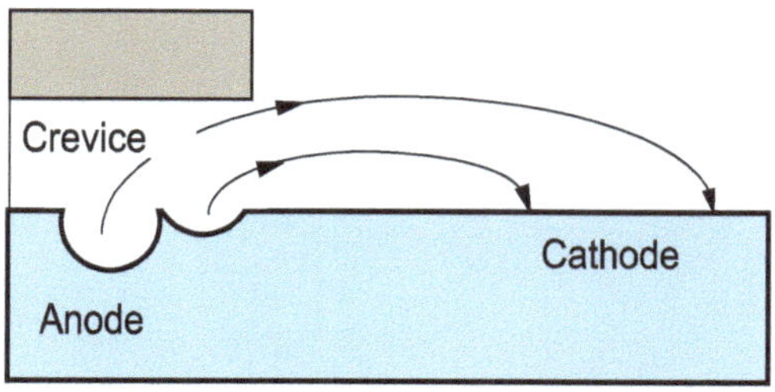

Figure 4.5 Scheme of crevice macrocell.

4.5 Appendix

4.5.1 Coefficient of variation, CV

As discussed in Chapter 8, Statistical Analysis of Corrosion Data, corrosion rates calculated by the algorithms have a stochastic meaning, because in practice a corrosion process occurs through a distribution of values rather than a fixed, constant, determined one.

The use of the calculated value, as well as for measured values in testing or in inspections, should take into account this interpretation. As said, CV derived from the experience can be suitably used for further calculations such as the maximum expected value.

When dealing with localized corrosion, in which macrocell mechanism applies, the following CV values can be adopted:

- for pitting potential, E_{pit}: CV = 0.4 (CV = 1 if following an exponential distribution as reported in literature [7] hence not applicable because of possible negative values);
- for PIT: CV = 0.4 (in some instances, 0.6 could also be considered). On the basis of Gumbel statistics $PIT_{max} = 2.25\ \mu$;
- for propagation corrosion rate, C_{rate}: CV = 0.2. By adopting extreme values statistics, maximum value by Gumbel statistics is $C_{rate,max} = \mu \cdot (1 + 3.14CV)$; therefore, $C_{rate,max} = 1.63\ \mu$.

Table 4.1 Summary of equations for the calculation of initiation time and propagation rate for pitting and crevice

	Equation	Parameters
Pitting potential (V SHE)	$E_{pit} \cong \frac{298}{T+273} \cdot \left(\frac{PIN}{25} + \log(1+v) - 0.25 \cdot \log\left[\frac{Cl}{36}+1\right] + \frac{pH-7}{25}\right)$	Cl (ppm) chloride content Cl_2 (ppm) chlorine content E_c (V) potential of cathodic process E_{Cl_2} (V) potential of cathodic process E_{O_2} (V) potential of cathodic process $E_{O_2/bio}$ (V) potential of cathodic process E_{pit} (V) pitting potential E_{rup} (V) breakdown potential of passive film inside crevice ΔE_C (V) driving voltage for crevice initiation ΔE_{pit} (V) driving voltage for pitting initiation ΔV (V) driving voltage i_L (A/m^2) oxygen limiting current density k (h) exper. constant (≈ 1) O_2 (ppm) oxygen content ρ (Ω m) resistivity PIN = PREN − 0.1%Mn − 100%S PREN = %Cr + 3.3%Mo + X%N $X = 16$ for austenitic $X = 30$ for duplex T (°C) temperature v (m/s) fluid velocity
Potential of cathodic processes E_c (V SHE)	$E_{O_2} \cong 1.23 - 0.33\log\left(\frac{50}{[O_2]}\right) - 0.059\ pH$ $E_{O_2/bio} = E_{O_2} + 0.2$ $E_{Cl_2} = 1.36 + 0.6\log\left[\frac{Cl_2}{36}\right]$	
Pitting equation, pitting induction time, PIT (h) or general pitting equation	$E_{pit} \cdot (T+273) \cong 298\log\left(2.5 \cdot \prod_i Y_i\right)$ $PIT \cong k\ 10^{\left(\frac{PREN}{2\log[Cl]}(1-\Delta E_C)\right)}$ $\Delta E = E_C - E_{pit}$	
Pitting and crevice critical chloride concentration (ppm)	$\log[Cl^-]_{critical} \cong \left[\frac{PIN}{9.5} + \frac{pH-7}{5} - \frac{T-25}{100}\right]$	
Crevice critical gap size (CCGS) (μm)	$CCGS \cong \left(\frac{17}{PREN}\right)^2$	
Crevice induction time, CIT (h)	$CIT \cong k\ 10^{\frac{PREN}{3\log[Cl]}(1-\Delta E_C)}$	
Breakdown potential of passive film, E_{rup} (V SHE)	$E_{rup} \cong \frac{298}{T+273} \cdot \left(\frac{PREN}{25} - 0.25 \cdot \log\left[\frac{Cl}{36}+1\right] - 0.20\right)$	
Macrocell corrosion rate, $C_{rate,MC}$ (mm/y)	$C_{rate,MC} \cong 20\sqrt{\frac{i_L \cdot \Delta V}{(\rho+0.8)}}$	
Pitting Corrosion Rate, $C_{rate,pit}$ (mm/y)	$C_{rate,pitting} \cong 20\sqrt{\frac{i_L}{(\rho+0.8)}}$	
Crevice corrosion rate, $C_{rate,crevice}$ (mm/y)	$C_{rate,crevice} \cong 10\sqrt{\frac{i_L}{(\rho+0.8)}}$	

Table 4.A1 Values of parameters used for validation of general pitting equation

PREN	Environment	Chlorine, Cl_2	PIT (h)
18	Seawater/biofilm		$<10^2$
18	Fresh water	Yes	$\approx 10^3$
25	Seawater/biofilm		$<10^3$
40	Seawater/biofilm		$\approx 10^4$
50	Seawater	Yes	$>10^4$

4.5.2 Parameters used to validate general pitting equation

General pitting equation (4.6) has been validated on the basis of the experience data summarized in Table 4.A1.

4.5.3 Accelerated testing – pitting

General Pitting equation (4.8) and pitting potential equation (4.4) allow to design accelerated tests to determine:

- PIT for given operating conditions (including applied potential);
- pitting potential;
- PREN threshold.

4.5.3.1 Extrapolation of PIT

With reference to Chapter 7, Testing, Eq. (4.8) is a case of general Eq. (7.1), in which an affecting parameter is embedded in the driving voltage, ΔE. By taking logs, Eq. (4.8) becomes:

$$\log \mathrm{PIT} \cong A - B[Y] \tag{4.19}$$

where A and B are constants and Y is the affecting parameter, for instance, [Cl], T, pH and others. By plotting at least three accelerated conditions with reference to the selected parameter, Y, as shown in Fig. 4.A1, it makes possible an extrapolation at operating conditions.

4.5.3.2 Extrapolation of pitting potential

In principle, to obtain the straight line representing the pitting equation, two testing conditions would be sufficient and at least three would allow a more reliable extrapolation. Pitting equation (4.7) can be generalized at constant temperature as follows:

$$E_{\mathrm{pit}} \cong A + \sum \log B_i[Y_i] \cong A + \log \prod B_i[Y_i] \tag{4.20}$$

where A and B_i are constants and Y_i is the ith affecting parameter. When studying the influence of a single parameter, for example chloride content, [Cl], and keeping constant all other parameters, Eq. (4.20) reduces to:

$$\log[\mathrm{Cl}] \cong C - B_{\mathrm{Cl}} E_{\mathrm{pit}} \tag{4.21}$$

where B_{Cl} and C are constants. Fig. 4.A2 depicts the experimental plot of Eq. (4.21) for determining pitting potential, E_{pit}, at expected operating conditions through accelerated test in chloride-enriched solutions.

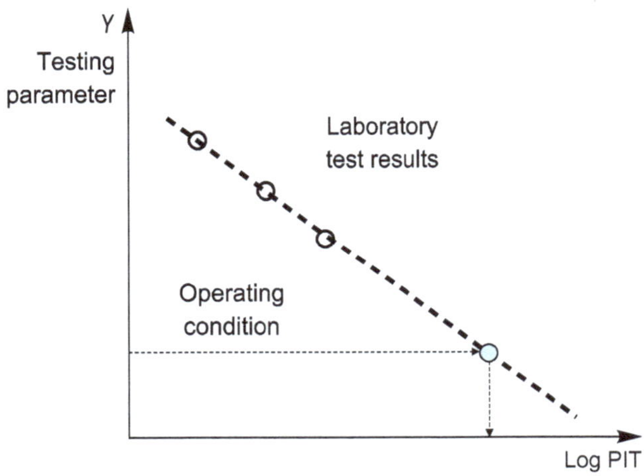

Figure 4.A1 Pitting equation plot for testing parameter Y.

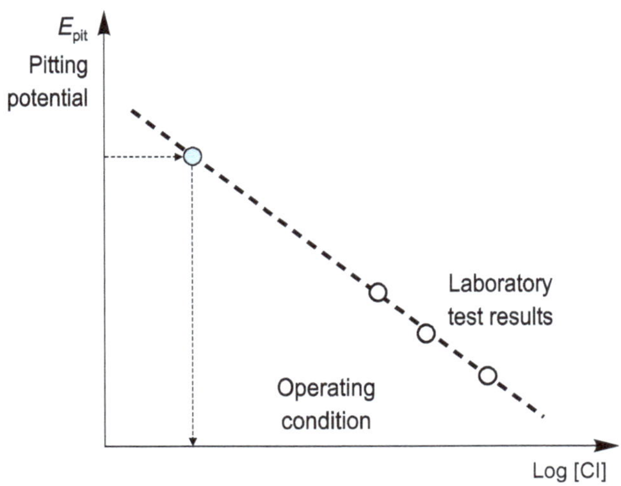

Figure 4.A2 Experimental plot for determining pitting potential at operating chloride concentration.

4.5.3.3 Extrapolation of minimum PREN

An equation similar to (4.21) can be derived for PREN (or PIN), as for instance the following:

$$\mathrm{PREN} \cong D - B_{\mathrm{PREN}} E_{\mathrm{pit}} \tag{4.22}$$

where B_{PREN} and D are constants.

4.5.3.4 Influence of temperature

In accelerated tests, to reduce testing time it is common practice to increase temperature, hence instead of varying one parameter, a couple of parameters is considered, for example chloride content and temperature as combined parameters. From pitting Eq. (4.7) by taking constant all parameters except a testing one, for instance chloride, in combination with temperature, Eq. (4.7) becomes:

$$\log[\mathrm{Cl}]_i \cong C^* - B_{\mathrm{Cl},T} \cdot [T \cdot E_{\mathrm{pit}}]_i \tag{4.23}$$

where index i refers to the testing condition (generally, more than two); $B_{\mathrm{Cl},T}$ and C^* are constants. Fig. 4.A3 depicts the relevant experimental plot of Eq. (4.23).

4.5.3.5 Example of extrapolation of pitting potential

To characterize a new stainless steel with PREN 30 for implant applications, pitting relating parameters are required for acceptance. Expected operating conditions are

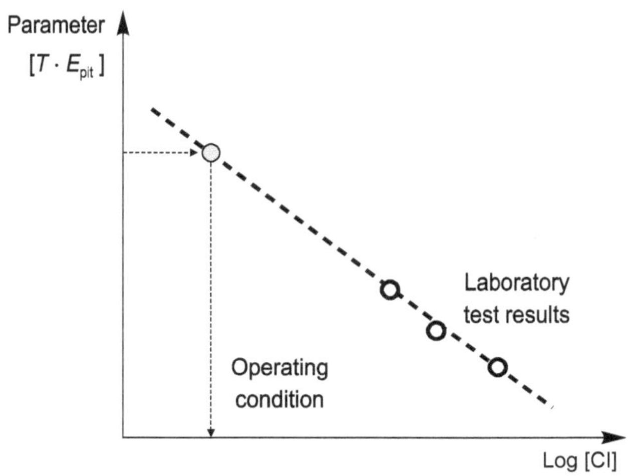

Figure 4.A3 Experimental plot for determination of pitting potential at operating temperature and chloride concentration.

the following: physiological solution with a chloride content of 0.9% NaCl equivalent ($[Cl^-] = 5400$ ppm) at body temperature (37°C) stagnant conditions.

Accelerated test is designed as follows:

- 10% NaCl – test temperature 67°C ($[Cl^-] = 61{,}000$ ppm);
- 5% NaCl – test temperature 57°C ($[Cl^-] = 30{,}500$ ppm);
- 2.5% NaCl – test temperature 47°C ($[Cl^-] = 15{,}250$ ppm).

Pitting measurements were carried out as follows:

- 9 specimens were prepared (same finishing as specified for applications, final polishing in dry wheat) and divided into three sets by random sampling;
- pitting potential of each specimen was measured by means of potentiodynamic potential scan with 10 mV–1 min step when anodic current reached 1 A/m^2;
- Mean of three specimens for each testing condition, resulted as follows:
  - $[(273 + 67){-}61{,}000]_1$; $E_{pit,1} = 110$ mV SCE;
  - $[(273 + 57){-}30{,}500]_2$; $E_{pit,2} = 250$ mV SCE;
  - $[(273 + 47){-}15{,}250]_3$; $E_{pit,3} = 390$ mV SCE.

Fig. 4.A4 shows the plot, $[T \cdot E_{pit}] - \log[Cl]$, according to the pitting Eq. (4.7), whose equation is $Y = -133[Cl] + 680$, where $Y = [T \cdot E_{pit}]/1000$. At expected operating conditions: ($[Cl^-] = 5400$ ppm, temperature 37°C, stagnant conditions) E_{pit} is 590 mV SCE. Stainless steel was accepted for applications.

4.5.4 Accelerated testing-crevice corrosion

Standards for testing crevice susceptibility, such as ASTM G78 [8], are proposed to rank alloys when exposed to a specified solution. A multiple crevice assembly is used, which consists of two nonmetallic segmented washers, each having a number of grooves and plateaus. Each plateau, in contact with the metal surface, provides a

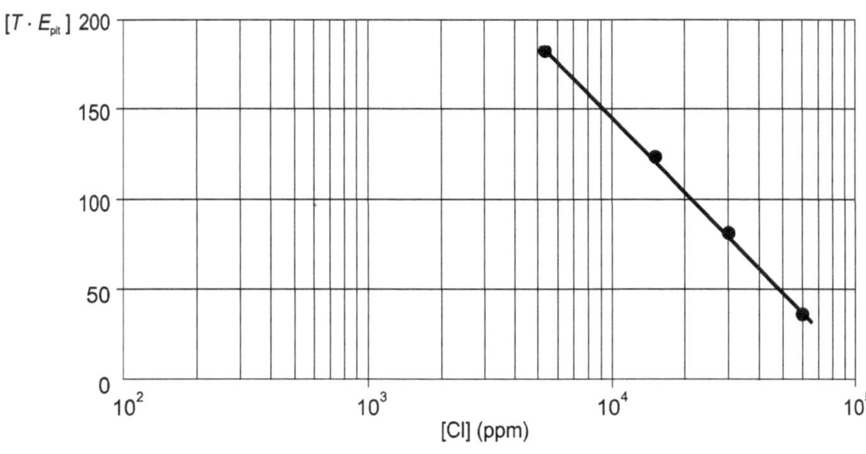

Figure 4.A4 Example of accelerate test results for Pitting Potential determination.

possible site for initiation of crevice corrosion. Parameters measured after the test are as follows:

- number of sites (grooves) where crevice occurred (as percentage);
- mass loss;
- depth of attack.

Interpretation of results seems to be arbitrary and meaningless for the following:

- if testing cathodic process is oxygen reduction, as it is in operating, it has to be realized that as soon as a crevice has started there is a cathodic protection effect on nearest sites, therefore the number of crevice sites, which would be shown, is strongly reduced;
- if noble cathodic process is used, as ferric ion containing solution, crevices start regardless the real operating condition, and once started, the number of activated sites become random. This test, rather than producing a rank, seems to be a pass–fail-test where most likely only a few high PREN alloys can pass;
- if cathodic process is less noble than oxygen reduction, as in deaerated acidic solutions, mass loss is much lower than that in operating with different cathodic process; conversely, number of active sites could be higher because of less pronounced cathodic protection effects;
- depth of crevice depends on the cathodic process occurring during the test and is independent from the alloy, as the macrocell mechanism is governed by the geometry of cell assembly, electrolyte conductivity and cathodic process, only.

In conclusion, tailored single site crevice specimen, with specific surface area ratio, should be used.

4.6 Case studies

In the following, some examples of the use of proposed algorithms are reported.

4.6.1 *Prediction of pitting occurrence based on PIT*

In designing heat exchanger tubes with seawater as cooling fluid, three candidate metals were considered, namely: (1) 18–8 stainless steel (AISI 304 grade); (2) 18-10-3 stainless steel (AISI 316 grade) and (3) high-alloy austenitic stainless steel with 6% Mo. The design water velocity was 1 m/s as first choice.

The acceptance criterion could be the PIT, with threshold limit 10^4 h (approximately 1 y): if PIT exceeded this threshold, the probability of pitting initiation would be taken as very low, hence the metal would be accepted.

To estimate PIT, the following parameters are used:

- seawater: $[Cl^-] = 19{,}000$ ppm; 8.3 pH;
- water velocity, 1, 1.5 and 2 m/s;
- oxygen content: 6 ppm;
- temperature: 35°C;
- chlorine (optional): 2 ppm (residual).

Table 4.A2 Results of calculation of pitting potential E_{pit} and pitting induction time, PIT

PIN	E_{pit}	E_{O_2}	E_{Cl_2}	$PIT_{(O_2)}$	$PIT_{(O_2-bio)}$	$PIT_{(Cl_2)}$
Water velocity 1 m/s						
	V SHE	V SHE	V SHE	h (≈)	h (≈)	h (≈)
17	0.341	0.436	0.607	***<100***	***<50***	***<50***
28	0.766			22,000	***5000***	***6500***
35	1.037			>100,000	>100,000	>100,000
45	1.424			>100,000	>100,000	>100,000
Water velocity 1.5 m/s						
17	0.435	0.436	0.607	***100***	***<50***	***<50***
28	0.860			45,000	***10,000***	12,000
35	1.131			>100,000	>100,000	>100,000
45	1.510			>100,000	>100,000	>100,000
Water velocity 2 m/s						
17	0.511	0.436	0.607	**150**	***600***	***700***
28	0.937			80,000	18,000	22,000
35	1.208			>100,000	>100,000	>100,000
45	1.595			>100,000	>100,000	>100,000
Stagnant condition						
17	0.050	0.436	0.607	***<20***	***<10***	***<10***
28	0.475			***2500***	***500***	***700***
35	0.746			>100,000	30,000	40,000
45	1.133			>100,000	>100,000	>100,000

Note: Values in bold-italics indicate certain pitting initiation condition.

Results of the application of equations proposed for the estimation of Pitting Potential E_{pit} and PIT, are reported in Table 4.A2. It can be concluded that AISI 304 grade is not suitable; AISI 316 grade is suitable for water velocity exceeding 1.5 m/s; higher grades resist at all conditions.

It is worth noting that in stagnant condition, only highest grade can resist providing chlorine injection to avoid formation of biofilm. This result has an impact on shutdown duration which should be limited to a few weeks.

4.6.2 Estimation of perforation time

A tank designed to store natural, i.e., not treated, seawater was made of stainless steel, 18–8 grade (AISI 304) with 4-mm-thick bottom plate. It suffered pitting and experienced perforation with leakage in short time. After the incident, an assessment was made as revision of the design to check:

- pitting conditions as PIT;
- perforation time;
- comparison between natural versus sterile seawater;
- alternative metals;
- feasibility of CP and protection potential interval.

Operating conditions were found as follows:

- $[Cl^-] = 19{,}000$ ppm;
- pH = 8.3;
- temperature = 15°C;
- oxygen content = 8 ppm;
- stagnant conditions.

To ascertain whether conditions for pitting initiation applied, PIT has to be estimated through Eq. (4.8): It can be considered that PIT below about 1000 h indicates the real incubation time before propagation would start.

Table 4.A3 reports results of the calculation of PIT and pitting potential at expected operation conditions.

Either grade 304 or grade 316 suffers pitting initiation within 1 day for 304 and 1 week for 316. Also sterile water would experience pitting initiation within a longer time, about 1 week and 1 month, respectively.

The perforation time depends on corrosion rate, which can be estimated through Eq. (4.17), based on the following data: seawater resistivity = 0.25 Ωm; oxygen limiting current density, $i_L \cong 80$ mA/m^2, as calculated by Eq. (1.35). Corrosion rate is about 5.6 mm/y.

Perforation time is 0.7 year that is about 9 months, including a week for the incubation time for nontreated seawater and 1 month longer, i.e. 10 months for 316, for sterile aerated seawater.

In conclusion, the material selected in the design phase was a wrong choice.

Table 4.A3 Calculation of pitting potential, E_{pit}, and pitting induction time, PIT

Stainless steel grade	PIN = PREN	E_{pit}	E_{O_2}	E_{O_2-bio}	$PIT_{(O_2)}$	$PIT_{(O_2-bio)}$
		V SHE	V SHE	V SHE	*h* (≈)	*h* (≈)
AISI 304	17	0.053	0.478	0.678	***14***	***5***
AISI 316	28	0.508	0.478	0.678	***2300***	***500***

Note: Values in bold-italics indicate certain pitting initiation condition.

As alternate stainless steels, based on expected PIT higher than 10^4 h, for non-treated seawater, assuming that biofilm may form, it seems that minimum PREN would be 45; for sterile seawater, even with 2 ppm of chlorine added, PREN 35 would resist pitting initiation as well as crevice.

A very good, valuable and cheap solution would be low grade stainless steel, as 304 grade, with CP, either by galvanic anode or an ICCP. Assuming to use AISI 304, protection potential would be at least 100 mV below the pitting potential, which is about 0.1 V SHE (corresponding to −0.2 V SCE) calculated by Eq. (4.4). As potential of an iron anode is −0.65 V SCE, CP employing iron anodes does work establishing protection condition by passivity [9]. In the case an ICCP is used, the potential interval to be fixed for checking protection condition could be between −0.3 and −0.5 V SCE.

4.6.3 A case study

A plate made of stainless steel, grade AISI 304 (18–8 Cr–Ni with PREN 17), immersed in a swimming pool, experienced pitting near welds. Before proceeding with remedial actions, it was considered mandatory to establish the most likely root cause of pitting, as well as the estimation of pitting initiation time and pitting propagation rate.

The water of the pool was fresh water from local public main, with the following characteristics:

pH = 7.5
Temperature: ≈ 27°C
TDS = 0.45 g/L (resistivity ≅ 20 Ωm)
Chloride = 55 mg/L
Sulphate = 134 mg/L
Oxygen (at saturation) = 8 ppm
Chlorine = 5–10 ppm.

In design, AISI 304 grade stainless steel was chosen on the basis of the CCC, calculated by Eq. (4.9) which gives the value 71 ppm. It was not considered the use of chlorine, injected as chlorine dioxide, for sanitary requirements.

As reported in Table 4.A4, which summarizes the results of Eqs (4.1)–(4.8) for estimation of pitting potential, E_{pit}, potential of the cathodic process for reduction

Table 4.A4 Calculation of expected pitting induction time, PIT, for different chlorine dosage

PIN ≅ PREN	Cl_2	E_{pit}	E_{O_2}	E_{Cl_2}	$PIT_{(O_2)}$	$PIT_{(Cl_2)}$
		V SHE	V SHE	V SHE	h (≈)	h (≈)
18 (AISI 304)	5	0.58	0.55	0.85	***5000***	***2500***
18 (AISI 304)	10	0.58	0.55	1.03	***5000***	***1500***
25 (AISI 316)	5	0.89	0.55	0.85	>10,000	>10,000
25 (AISI 316)	10	0.89	0.55	1.03	>10,000	>10,000

of oxygen, E_{O_2}, and chlorine, E_{Cl_2}, and PIT, pitting could not occur because of the presence of oxygen, instead it could for the presence of a chlorine concentration exceeding 5 ppm, as used in swimming pools. It is worth noting that a higher grade as AISI 316, with PREN 28, should have resisted pitting.

Pitting started near welds where passive film is weaker and more susceptible to localize breakdown for the action of chloride ions. Once pitting started, corrosion rate follows Eq. (4.17); by inputting oxygen limiting current density, $i_L \cong 140\ mA/m^2$, as calculated by Eq. (1.35), water resistivity $\rho = 20\ \Omega m$, corrosion rate is about 1.7 mm/y.

In summary, pitting started after an incubation time of about 2 months preferentially near welds, where the passive film is weaker and proceeded at a corrosion rate of 1.7 mm/y.

4.6.4 Unexpected pitting (typically)

There is a typical although unexpected pitting occurrence in some industrial applications, as for example in chemical plants operating by a batch process. This often reoccurring case study can be exemplified as follows:

- a process is by batch involving a chloride containing solution with a chloride content relatively high, for instance a few per cent by weight; metal selected for plant components was stainless steel, AISI 316L grade;
- after each batch, which lasts 48 h or less, acid cleaning, neutralization and repassivation procedures are carried out;
- a new plant section, as an extension of an existing one, experienced perforation by pitting attack after a few months from start-up;
- old section never exhibited pitting, although operating the same solution at same conditions.

After the astonishment, the question one asks is: how could it happen?

Let's start analysing the pitting occurrence condition for the old section. The first step is the evaluation of the PIT parameter on the basis of the following conditions: $[Cl] \approx 10{,}000$ ppm; $T = 50°C$; $pH \approx 7$; oxygen present. From Eq. (4.4), pitting potential is calculated: 0.27 V SHE; from Eq. (4.1), cathodic potential is calculated: 0.51 V SHE; from Eq. (4.8), PIT is calculated: ≈ 1100 h. Incubation time for pitting initiation is of the order of a month.

Based on the above result, it appears evident that pitting never starts because the batch time is much less than a month and besides this, repassivation is performed at each batch, hence zeroing the countdown of PIT.

When designing and building up the new section as equal to the existing one, no pitting could have been expected on the basis of the experienced operating. So, what new for the new section?

The answer is hydraulic testing before operating start-up. As unfortunately often it happens, although water used for the hydrotest was proper (i.e. low salinity, low chloride content, even deoxygenated) the operation of water discharge was faulty, because it left inside the plant some residual water, which concentrated by evaporation then allowing locally chloride concentration to reach critical values to trigger pitting within the time comprised between water discharging and the operating start-up. As it is known, once pitting started, it cannot be stopped: in other words, when operating of new section started, dormant pitting resumed inevitably.

References

[1] Pedeferri P. Corrosione e protezione dei materiali metallici. Italy: Milan: Polipress; 2007. ISBN 97888-7398-032-2 (in Italian).

[2] Pedeferri P. Cathodic protection and cathodic prevention, construction and building materials. Vol.10, n. 5, pp 391–402, 1996.

[3] Defranoux JM. Sur la résistance à la corrosion des aciers inoxydables dans l'eau de mer. France; Cannes: First International Congress on Marine Corrosion; 1964. in French.

[4] Mollica A, Trevis A. The influence of the microbiological film on stainless steels in natural seawater. In: Proc. of fourth international congress on marine corrosion and fouling, paper n. 351. France, Juan-les Pins; 1976.

[5] Oldfield JW. Test techniques for pitting and crevice corrosion resistance of stainless steels and nickel alloys in chloride containing environments. Int Mater Rev 1987;32 (3):153–70.

[6] Oldfield JW, Sutton WJ. New technique for predicting the performance of stainless steels in sea water and other chloride containing environments. Br Corr J, 15 1980;1:31–4.

[7] Kowaka M, Tsuge H, Akashi M, Matsumura K, Ishimoto H. An introduction to life prediction of industrial plant materials – application of extreme value statistical method for corrosion analysis. New York: Allerton Press; 1994.

[8] ASTM. G78-01 standard guide for crevice corrosion testing of iron-base and nickel-base stainless alloys in seawater and other chloride-containing aqueous environments.

[9] Lazzari L, Pedeferri P. Cathodic protection. Italy; Milan: Polipress; 2005. ISBN 88-7398-020-1.

Corrosion in Water, Soil and Air

5

Natural and industrial waters as well as some soils are corrosive toward carbon and low alloy steels if a cathodic process can take place. Condensed water from humid atmosphere can be corrosive because highly oxygenated and for the presence of pollutants such as chlorides and sulphur oxides. Oxygen reduction is the most typical cathodic process; however, hydrogen evolution in slightly acidic conditions and reduction of oxidizing species as chlorine are often present. More rarely, microbiologically induced corrosion, MIC, is present. An important factor that reduces aggressiveness is the tendency to form protective scales.

5.1 Freshwater

pH of freshwater is close to neutrality; hence, carbon and low alloy steels suffer corrosion if oxygen or other oxidizing species are present. In the absence of oxygen, MIC may take place by the action of anaerobic bacteria, particularly by the sulphate-reducing bacteria (SRB).

Corrosion morphology and corrosion rate depend on oxygen content and hydrodynamics conditions. Mild steel shows frequently general corrosion with some tendency to localize on small areas; copper, rarely subjected to corrosion, suffers localized attacks of pitting type. Recent standards proposed a corrosiveness classification of freshwater as probability of corrosion: negligible, low, moderate, severe and very severe. In the following, an engineering approach for the evaluation of the corrosion rate for carbon and low alloy steels is discussed.

5.1.1 Uniform corrosion in freshwater

In freshwaters, because of the presence of oxygen and chlorine, carbon steel suffers corrosion whose rate is governed by the diffusion laws, as reported in Chapter 2, Uniform Corrosion. Diffusion depends on oxygen and chlorine contents, temperature, turbulence and the presence of calcium carbonate scale if water has a scaling tendency. The formation of the calcium carbonate scale is governed by the chemical equilibrium of carbonate/bicarbonate (calcium carbonate is insoluble, whereas calcium bicarbonate is highly soluble). Scaling tendency is forecasted by indexes, which the most used is Langelier Saturation Index, LSI, calculated as follows.

$$\mathrm{LSI} = \mathrm{pH} - \mathrm{pHs} \tag{5.1}$$

Engineering Tools for Corrosion. DOI: http://dx.doi.org/10.1016/B978-0-08-102424-9.00005-7

where pH is the actual water pH, and pHs is *saturation* pH. Positive values of LSI indicate tendency to scaling and, conversely, negative values indicate tendency to solubilize the calcium carbonate deposits.

Saturation pHs is obtained from carbonate/bicarbonate equilibrium, as follows:

$$\text{pHs} = \log K_2 - \log K_{ps} - \log[\text{Ca}^{2+}] - \text{logalkM} \tag{5.2}$$

where K_2 is the second dissociation constant of carbonic acid, H_2CO_3; K_{ps} is the solubility product of $CaCO_3$; $[Ca^{2+}]$ is the concentration of calcium ions in mol/L; alkM is the total alkalinity in eq/L.

The following empirical relationships are used:

$$\begin{aligned} T < 25°\text{C pHs} = 12.65 - 0.0142(1.8T + 32) - \log[\text{Ca}^{2+}] \\ - \log(\text{alkM}) + 0.1 \log(\text{TDS}) \end{aligned} \tag{5.3}$$

$$\begin{aligned} T > 25°\text{C pHs} = 12.27 - 0.00915(1.8T + 32) \\ - \log[\text{Ca}^{2+}] - \log(\text{alkM}) + 0.1 \log(\text{TDS}) \end{aligned} \tag{5.4}$$

T is °C; $[Ca^{2+}]$ mg/L; alkM mg/L as $CaCO_3$; TDS mg/L.

For *not scaling waters*, defined by a negative value of LSI < 0, corrosion rate is derived from Eq. (2.13):

$$C_{\text{rate,water,LSI}<0} \cong 12 \cdot \left\{[\text{O}_2] + 0.04 \cdot [\text{Cl}_2]\right\} \cdot \left(1 + \sqrt{v}\right) \cdot 2^{\frac{T-25}{25}} \tag{5.5}$$

where symbols are known. $C_{\text{rate,water,LSI}<0}$ is in μm/y, $[O_2]$ and $[Cl_2]$ are, respectively, oxygen and chlorine content in ppm.

For *scaling waters*, LSI > 0, corrosion rate is determined by the permeability or porosity, p, defined as fraction of bare metal surface (i.e. $1 - p$ is scaling efficiency). This parameter was introduced by Bianchi [1] who noted that in the presence of a calcareous deposit the diffusion layer thickness coincides with scale thickness, so accordingly water velocity exerts no influence. Porosity, p, is a function of LSI as follows:

$$p \cong \frac{\left[10 - (\text{LSI})^2\right]}{10} \tag{5.6}$$

Therefore, corrosion rate, C_{rate}, in aerated freshwater with scaling tendency is given by:

$$C_{\text{rate,water,LSI}>0} \cong 1.2 \cdot \left\{[\text{O}_2] + 0.04 \cdot [\text{Cl}_2]\right\} \cdot \left[10 - (\text{LSI})^2\right] \cdot 2^{\frac{T-25}{25}} \tag{5.7}$$

where symbols are known.

5.1.2 Differential aeration in freshwater

In the case of *differential aeration*, corrosion rate, C_{rate}, is determined by the macrocell current between anodic corroding area and surrounding cathodic area, as seen before in Chapter 3, Localized Corrosion, Eq. (3.19), to give:

$$C_{rate,DA} \cong 20\sqrt{\frac{i_L \cdot \Delta V}{\rho}} \tag{5.8}$$

where symbols are known ($C_{rate,DA}$, mm/y; i_L, A/m^2; ΔV, V; ρ, Ωm, > 1 Ωm). By introducing the maximum driving voltage of 0.2 V, Eq. (5.8) shortens to:

$$C_{rate,DA} \cong 9\sqrt{\frac{i_L}{\rho}} \tag{5.9}$$

By introducing resistivity as function of salt concentration (TDS, g/L) and temperature (ΔT is $T - 25°C$):

$$\rho \cong \frac{9}{(1 + 0.02\Delta T) \cdot \mathrm{TDS}} \tag{5.10}$$

and oxygen limiting current density for scaling waters (LSI > 0), the following expression can be used:

$$C_{rate,DA,FW} \cong 0.1 \cdot \sqrt{[O_2] \cdot 2^{\frac{T-25}{25}} \cdot \left[10 - (\mathrm{LSI})^2\right] \cdot \mathrm{TDS} \cdot (1 + 0.02\Delta T)} \tag{5.11}$$

where symbols are known ($C_{rate,DA,FW}$ in mm/y; $[O_2]$, ppm; TDS, g/L; ρ, Ωm).

5.1.3 Galvanic corrosion in freshwater

In the presence of noble metals (i.e. copper, stainless steel) or noble oxides, typically magnetite and calamine, galvanic corrosion occurs. The general equation for the calculation of the corrosion rate is the same as Eq. (5.8):

$$C_{rate,GC} \cong 20\sqrt{\frac{i_L \cdot \Delta V}{\rho}} \tag{5.12}$$

where symbols are known (C_{rate} in mm/y). By introducing the driving voltage of maximum 0.5 V and the expression of oxygen limiting current density and resistivity as function of salt concentration, the following expression can be used:

$$C_{rate,GC,FW} \cong 0.15 \cdot \sqrt{[O_2] \cdot 2^{\frac{T-25}{25}} \cdot \left[10 - (\mathrm{LSI})^2\right] \cdot \mathrm{TDS} \cdot (1 + 0.02\Delta T)} \tag{5.13a}$$

where symbols are known. ($C_{rate,GC,FW}$ is in mm/y; $[O_2]$ is the oxygen content, ppm; TDS is the total dissolved solids or salinity, g/L; ρ is the freshwater resistivity, Ωm; ΔT is the temperature increase, °C).

For nonscaling freshwaters (LSI $<$ 0), influence of water velocity has to be considered; hence, Eq.(5.13a) becomes:

$$C_{rate,GC,FW} \cong 0.5 \cdot \sqrt{[O_2] \cdot 2^{\frac{T-25}{25}} \cdot \left(1+\sqrt{v}\right) \cdot TDS \cdot (1+0.02\Delta T)} \quad (5.13b)$$

where symbols are known ($C_{rate,GC,FW}$ in mm/y).

5.2 Seawater

Corrosion behaviour of carbon and low alloy steels continuously immersed in seawater depends on temperature and turbulence; in stagnant condition, corrosion rate decreases with time because corrosion products mixed with calcareous deposit (i.e. calcium carbonate) contribute to reduce oxygen availability. In summary, in stagnant conditions, affecting parameters are oxygen content, temperature and exposure time, whereas in highly turbulent conditions, they are oxygen content and water velocity; accordingly, the following equations can be used to estimate uniform corrosion rate of carbon and low alloy steels:

$$C_{rate,SW,stagnant} \cong \frac{12 \cdot [O_2] \cdot 2^{\frac{T-25}{25}}}{\sqrt{t}} \quad (5.14)$$

$$C_{rate,SW,flowing} \cong 12 \cdot [O_2] \cdot \left(1+\sqrt{v}\right) \cdot 2^{\frac{T-25}{25}} \quad (5.15)$$

where symbols are known. $C_{rate,SW}$ is in μm/y and t is in y, for $t > 1$ y.

5.2.1 Galvanic corrosion in seawater

In the presence of galvanic couplings with noble metals (i.e. copper, stainless steel) or noble oxides, typically magnetite, galvanic corrosion rate is given by the equation derived from Eqs (3.8) and (3.17) by rounding resistivity to 1 Ωm and driving voltage to 0.5 V:

$$C_{rate,GC,SW} \cong 1.4 \cdot \sqrt{[O_2] \cdot 2^{\frac{T-25}{25}} \cdot \left(1+\sqrt{v}\right)} \quad (5.16)$$

where $C_{rate,GC,SW}$ is in mm/y, $[O_2]$ is the oxygen content (ppm), v is the water velocity (m/s), and T is the temperature in °C. Eq.(5.16) gives maximum expected values in most typical applications.

5.3 Flow-enhanced corrosion

In flowing water, corrosion rate of active metals increases as water velocity increases and becomes catastrophic above a threshold, called *critical velocity*, because a mechanical erosion of surface layers (protective film or corrosion products) occurs. This type of corrosion attack is called *erosion-corrosion*.

The onset of erosion-corrosion depends on the synergy between the mechanical action exerted by the fluid and corrosion resistance properties of the metal. An empirical relationship proposed by API (American Petroleum Institute) [2] provides a link between the mechanical action exerted through the kinetic energy of the fluid and material property as follows:

$$v_{E-C} \cong \frac{C}{\sqrt{\gamma}} \tag{5.17}$$

where v_{E-C} (m/s) is the fluid critical velocity, C is a constant depending on metal, and γ (kg/m^3) is the fluid density. The meaning of the critical velocity is evident: As soon as fluid velocity exceeds the critical value ($v > v_{E-C}$), corrosion rate increases of at least an order of magnitude. Although this relationship was proposed for two-phase fluids, it can be extended to a single phase as water. As water density is close to 1000 kg/m^3, Eq. (5.17) shorts to:

$$v_{E-C} \cong 3 \cdot k_M \tag{5.18}$$

Metal constant k_M assumes the following values:

- Cu 0.4
- Cu–Ni (90/10) 0.9
- Cu–Ni (70/30) 1.2
- Mild steel (C–Mn steel) 1.2
- Low alloy steel (Cr–Mo steel) 1.5
- Monel 3
- Stainless steel 4

5.4 Soil

Uniform corrosion rate in soil is generally negligible because oxygen content is a few ppm. When conditions for localized corrosion exist, corrosion rate is no more negligible.

5.4.1 Differential aeration in soil

This type of corrosion, called *differential aeration*, occurs in nonhomogeneous soil, typically in clayey soil, less permeable to oxygen, adjacent to aerated sandy soil zones, as shown in Fig. 3.8. The alkalinity produced by the oxygen reduction passivates the steel as Fig. 3.10 shows, then stabilizing the macrocell. Corrosion rate is determined by the macrocell current between anodic corroding area (clayey soil) and surrounding cathodic area (sandy soil). From Eqs (1.35) and (3.8) by introducing a maximum driving voltage of 0.2 V, expected corrosion rate through a differential aeration mechanism is obtained:

$$C_{\text{rate,DA,soil}} \cong \sqrt{\frac{[O_2]}{\rho}} \tag{5.19}$$

where oxygen concentration in the one of aerated zones (in ppm), ρ is the soil resistivity (in Ωm), and $C_{\text{rate,DA,soil}}$ is in mm/y.

5.4.2 Galvanic corrosion in soil

In the presence of a galvanic coupling, corrosion rate of carbon and low alloy steels coupled with more noble metals as copper and stainless steel can be derived from Eqs (1.35) and (3.8) by introducing a maximum driving voltage of 0.5 V, as follows:

$$C_{\text{rate,GC,soil}} \cong 1.5\sqrt{\frac{[O_2]}{\rho}} \tag{5.20}$$

where $C_{\text{rate,GC,soil}}$ is the maximum expected galvanic corrosion rate in mm/y; $[O_2]$ (ppm) is the oxygen concentration in cathodic zones, and ρ (Ωm) is the soil resistivity.

5.5 Microbiologically induced corrosion

Corrosion caused by microorganisms, called MIC is often encountered in several industrial plants: production, transport and storage of hydrocarbons; fire-fighting systems; water cooling circuits; sewage treatment plants; beneath fouling on marine structures; buried structures in clayey and swampy soils.

First step of this corrosion process is the formation of so-called *biofilm*, which consists of microorganisms colonies, stuck on metal surface by self-produced gels, which modify locally pH and oxygen availability and create conditions for bacteria thriving.

Second step occurs as bacteria grow and consists on the setup of electrochemical conditions for the corrosion process, namely the establishment of a cathodic process that must be nobler than the anodic one.

5.5.1 Cathodic processes promoted by bacteria

In short, bacteria can promote three cathodic processes, namely:

- hydrogen evolution (i.e. acidic corrosion) in either aerated or deaerated (anaerobic) environments by formation of sulphuric or organic acids, as acetic, formic, pyruvic and others;
- reduction of sulphate to sulphide in anaerobic environment. In sterile environments, sulphate reduction reaction does not take place for kinetics reasons or in other words, because there is no catalysers to trigger the reaction;
- oxygen reduction as consequence of start-up of localized corrosion on passive metals as stainless steels. This occurrence involves pitting corrosion of stainless steels in aerated freshwater or seawater.

5.5.1.1 Acid-producing bacteria

There are bacteria, typically *aerobic* as sulphur-oxidant, producing sulphuric acid, and *anaerobic*, which produce, through their metabolism, some organic acids. When producing sulphuric acid, corrosion rate is directly related to the pH (see Section 2.1.1, stagnant conditions), whereas when organic acids are produced, corrosion rate is determined by the formation of complex as discussed in Sections 2.1.4 and 2.4.3. Examples are *aerobic bacteria* are *Thiobacillus thiooxydans*, *Thiobacillus ferroxidans* and *Gallionella*.

5.5.1.2 Sulphate-reducing bacteria

Corrosion of carbon and low alloy steels in anaerobic, near-neutral environments is at first sight quite surprising. However, as said, it occurs through a cathodic process consisting of the reduction of sulphate to sulphide. This cathodic reaction, although as noble as oxygen reduction in neutral solutions, cannot take place spontaneously, if not catalysed, because of its slow kinetics. The most known catalyser is given by bacteria metabolism, as proposed by Von Wolzogen and Van der Vlugt [3]. In short, an enzyme, *hydrogenase*, allows the reduction of sulphate to sulphide: $SO_4^{2-} + 8H^+ + 8e^- \rightarrow S^{2-} + 4H_2O$. Indeed, this reaction occurs also in nature, in anaerobic, sulphate-enriched environments, where the oxidation reaction is carbon to carbon dioxide or methane; if these bacteria find metallic iron available, the thermodynamically more preferred anodic reaction becomes iron oxidation.

Recognized *anaerobic bacterium* of this type is *Desulfovibrio desulfuricans* also called sulphate-reducing bacteria, SRB, which drives above mechanism.

5.5.1.3 Biofilm-forming bacteria

Stainless steels may suffer pitting corrosion in seawater or chloride containing waters in aerobic condition when contaminated by bacteria, because a *biofilm* forms that rises the potential of about 200–300 mV compared to sterile equivalent sodium chloride containing solution. This ennoblement of the potential triggers the pitting initiation if pitting potential is exceeded ($E > E_{pit}$, refer to Chapter 4: Pitting and

Crevice Corrosion). Nevertheless, pitting on stainless steels by the presence of a biofilm, which causes an ennoblement of the potential of the oxygen reduction reaction, is not always strictly considered a MIC phenomenon.

In freshwater, what rises the potential is the presence of aerobic bacteria of strain classes: *Leptothrix*, *Siderocapsa*, *Pedomicrobium* and *Hyphomicrobium* [4,5]. *Leptospirillum oxidans* is the most common bacterium, called MOB (*Manganese Oxidizing Bacteria*) which produces the manganese couple, Mn^{3+}/Mn^{2+}, having a standard potential as noble as +1.51 V SHE sufficient to trigger pitting initiation. A manganese concentration as low as 2 μg/L and a chloride content of only 40 mg/L are sufficient for that bacterium to initiate pitting of 18/8 stainless steel, as found in many European rivers (Rhine, Seine, Danube and others) [5].

5.5.2 A unified model for MIC

Whatever is the cathodic process, MIC is definitively an electrochemical related corrosion process as it stops by application of cathodic protection.

Possible cathodic processes are reported above, all complying the electrochemical requirement of being nobler than the anodic one. Table 5.1 summarizes the MIC mechanism for each cathodic process: acidic or organic acid; anaerobic by SRB; ennoblement by biofilm-forming bacteria.

5.5.3 Corrosion rate on mild steel by SRB

Corrosion rate coincides with the rate of the cathodic reaction, which is sulphate to sulphide reduction, catalysed by an enzyme, hydrogenase, that SRB releases:

$$SO_4^{=} + 8H^+ + 8e = 4H_2O + S^{=} \quad (5.21)$$

The rate of that cathodic reaction is regulated by overvoltage; in this reaction, the prevailing overvoltage is concentration polarization of $SO_4^{=}$ from the bulk to the tuberculum, where SRB have settled for thriving. Inside the tuberculum, sulphates deplete to give sulphide ions, which precipitate as iron sulphide, so concentration of sulphate ions on the corroding metal surface zeros. It can be assumed that overvoltage is effectively given by the concentration polarization, leading to a diffusion control. On the other hand, experience has shown since long time that corrosion rate maximizes around 40°C (i.e. the physiological temperature) and is about 1.2 mm/y in seawater or in the presence of a sulphate content of about 2000 ppm, and almost symmetrically decreases below and over this temperature. Accordingly, based on that, the maximum expected corrosion rate depends on sulphate content and temperature, following an empirical relationship as:

$$C_{\text{rate,SRB}} \cong [SO_4^{2=}]^{0.5} - 0.0016 \cdot (T-38)^2 \quad (5.22)$$

Table 5.1 MIC mechanisms

Metal/ bacteria	MIC mechanism	Key parameters
Mild steel *Sulphuric acid-producing bacteria* (thiooxydant)	Anodic process: $Fe = Fe^{2+} + 2e$ Cathodic process: $2H^+ + 2e = H_2$	pH
Mild steel *Organic acid-producing bacteria* (gallionella)	Anodic process: $Fe = [org\text{-}Fe^{2+}]_{complex} + 2e$ Cathodic process: $2H^+ + 2e = H_2$	Complex constant
Mild steel *Sulphate-reducing bacteria* (SRB)	Anodic process: $Fe = Fe^{2+} + 2e$; $Fe^{2+} + S^{=} = FeS$ Cathodic process: $SO_4^{=} + 8e + 8H^+ = S^{=} + 4H_2O$	Sulphate content, temperature
Stainless steel (pitting) *Manganese oxidizing bacteria* (MOB) in freshwater	Anodic process: $M = M^{2+} + 2e$ Cathodic process: $O_2 + 2H_2O + 4e = 4OH^-$ MOB mechanism for potential ennoblement $4Mn^{2+} + 2H_2O + O_2 = 4Mn^{3+} + 4OH^-$ $4Mn^{3+}\ 4e = 4Mn^{2+}$	$[Mn^{2+}]$ and [Cl] content
Stainless steel (pitting) *Biofilm producing bacteria* in seawater	Anodic process: $M = M^{2+} + 2e$ Cathodic process: $O_2 + 2H_2O + 4e = 4OH^-$	Potential ennoblement

where $C_{rate,SRB}$ (mm/y) is the maximum corrosion rate, $[SO_4^=]$ (g/L) is the sulphate concentration and T (°C) is the temperature. Eq. (5.22) applies for a sulphate concentration in the range 50–2000 ppm (0.05–2 g/L) and temperature between 10 and 70°C. Negative values indicate no SRB attacks. In practice, corrosion rate hardly exceeds 1 mm/y, although a high sulphate concentration is present because iron sulphide works as protection layer.

5.6 Corrosive atmospheres

When dew forms from a humid atmosphere on a metal surface a corrosion attack can take place due to the presence of oxygen. Contaminants, as chloride and sulphur oxides, may accelerate the corrosion rate.

To establish whether corrosion conditions apply, two main parameters are important:

- *Relative humidity (RH)* which determines when condensed water forms;
- *Wetting factor (WF)* which determines the fraction of time when dew forms so corrosion attack takes place.

The following parameters influence corrosion rate:

- temperature of metal surface (generally coincident with the atmosphere temperature);
- presence of chloride and sulphate ions on the metal surface which has a twofold effect: makes active the metal and lowers the resistivity of the condensed water.

5.6.1 Conditions for corrosion occurrence

Condensed water, formed as dew, originates by four condensation mechanisms:

- *Physical*: it occurs when the temperature of metal surface is below ambient temperature;
- *Adsorption*: as result of physical attraction between metal and water molecules. It occurs even at low RH;
- *Chemical* due to hydroscopic effects caused by the presence on metal surface of salts; also corrosion products show this effect. It occurs even at very low RH;
- *Capillary*: it occurs on rough surfaces covered by patinas or corrosion products. Like the chemical one, it occurs even at very low RH.

In conclusion, condensation occurs frequently, for various mechanisms, unless RH is very low, for instance much lower than 40%. The thickness of the film varies from 1 to 10 μm for RH approaching 100%.

5.6.2 Corrosion model of atmospheric corrosion

Atmospheric corrosion occurs through the following features:

- formation of a film of condensed water; oxygen saturated with pH near 5.5 due to the presence of carbon dioxide in the atmosphere (actual content 400 ppm); RH is the governing parameter;

- corrosion stops when condensed film evaporates;
- corrosion occurrence is governed by WF parameter;
- *cathodic process* is primarily oxygen reduction, occurring without diffusion control;
- *overall corrosion factors* are RH, WF, temperature and conductivity of the film; the last depends on salt dissolved in the film.

Total thickness loss, Δs, is then given by the following equation:

$$\Delta s \cong \sum_{i=1}^{DL} \mathrm{WF}_i \cdot C_{\mathrm{rate},i} \cdot \Delta t_i \tag{5.23}$$

where Δt_i (y) is the homogeneous period of time at which $C_{\mathrm{rate},i}$ (μm/y) corrosion rate and WF_i wetting factor apply; DL (y) is the design life or total exposure time.

5.6.2.1 Time-dependence of corrosion rate

With reference to [6], results of testing and experience have proved that corrosion rate decreases with time, following a general trend given by a parabolic law:

$$C_{\mathrm{rate,atm}} \cong \frac{C_{0,\mathrm{atm}}}{\sqrt{t}} \tag{5.24}$$

where $C_{\mathrm{rate,atm}}$ (μm/y) is the corrosion rate at time t, $C_{0,\mathrm{atm}}$ (μm/y) is the initial corrosion rate, and t (y) is the time for $t > 1$ y. The reason of such behaviour derives from the barrier effect given by the corrosion products.

5.6.2.2 Initial corrosion rate

Once water condensation happened (RH $\approx$ 100%) and WF is unitary, initial corrosion rate, $C_{0,\mathrm{atm}}$, is function of temperature and electrolyte conductivity through an equation, which is as follows:

$$C_{0,\mathrm{atm}} \cong k_{\mathrm{atm}} \cdot [100 + (\mathrm{Cl} + \mathrm{SO_2})] \cdot 2^{\frac{T-25}{25}} \tag{5.25}$$

where k_{atm} is a constant which depends on metal as follows:

- $k_{\mathrm{atm,Fe}}$ 1
- $k_{\mathrm{atm,CorTen}}$ 0.5
- $k_{\mathrm{atm,Zn}}$ 0.04
- $k_{\mathrm{atm,Cu}}$ 0.02

$C_{0,\mathrm{atm}}$ (μm/y) is the initial corrosion rate, T (°C) is the temperature from 0 to 75, $[\mathrm{Cl} + \mathrm{SO_2}]$ (μg/m^3) is the sum of chloride and sulphur dioxide in the atmosphere.

Eq. (5.25) gives results in accordance with ISO 9223 [6] which proposes an equation derived as regression from laboratory exposure testing results. The rational of Eq. (5.25) is that temperature affects the kinetics of the oxygen reduction (reaction rate doubles every 25°C increase), whereas the presence of soluble salts (chlorides and sulphates) increases conductivity [reference is made to Eq. (5.10)] hence allowing an acceleration of the kinetics; constant, k_{atm}, reflects the behaviour of metal: active for iron and mild steel and passivating for others.

5.7 Summary

Table 5.2 summarizes equations for the calculation of corrosion rate in natural environments, namely water, soil and atmosphere.

5.8 Appendix

5.8.1 Coefficient of variation, CV

As discussed in Chapter 8, Statistical Analysis of Corrosion Data, corrosion rates calculated by the algorithms have a stochastic meaning, because in practice a corrosion process occurs through a distribution of values rather than a fixed, constant, determined one.

The use of the calculated value, as well as for measured values in testing or in inspections, should take into account this interpretation. As said, CV derived from the experience can be suitably used for further calculations such as the maximum expected value.

When dealing with general and localized corrosion in waters, soil and atmosphere, the following CV values can be adopted:

- *General corrosion*: CV = 0.2. The maximum expected value can be taken as $\mu + 2\sigma$ (normal distribution), where the standard deviation, σ, is obtained from the coefficient of variation, CV. By assuming a CV equals to 0.2, the maximum expected value is $x_{max} \cong 1.4\ \mu$;
- *Galvanic and differential aeration corrosion*: CV = 0.4 (in some instance, 0.6 could also be considered). The maximum expected value can be taken as $\mu + 2\sigma$, where the standard deviation, σ, is obtained from the coefficient of variation, CV. By assuming a CV equals to 0.4, the maximum value is $x_{max} \cong 1.8\ \mu$. On the basis of Gumbel statistics $x_{max} = 2.25\ \mu$;
- *MIC*: CV = 0.1. By adopting extreme values statistics, maximum value by Gumbel statistics is $x_{max} = \mu \cdot (1 + 3.14\ \text{CV})$; therefore, $x_{max} = 1.3\ \mu$;
- *Atmospheric corrosion*: same as general corrosion.

Table 5.2 Equations for the calculation of corrosion rate in natural environments

	Equation	Parameters
Uniform corrosion rate in NON scaling freshwater (μm/y)	$C_{\text{rate,water,LSI}<0} \cong 12 \cdot \{[O_2] + 0.04 \cdot [Cl_2]\} \cdot (1 + \sqrt{v}) \cdot 2^{\frac{T-25}{25}}$	alkM (mg/L as $CaCO_3$) $[Ca^{2+}]$ (mg/L) $[Cl_2]$ (ppm) chlorine content ΔT is $T - 25°C$ ΔV (V) driving voltage LSI Langelier saturation index $[O_2]$ (ppm) oxygen content ρ (Ωm) resistivity (>1 Ωm) v (m/s) water velocity T (°C) temperature TDS (mg/L)
Saturation Index or Langelier Saturation Index, LSI	$LSI = pH - pHs$ $T < 25°C{:}pHs = 12.65 - 0.0142(1.8T + 32) - \log[Ca2^{+}] - \log(\text{alkM}) + 0.1\log(\text{TDS})$ $T > 25°C{:}pHs = 12.27 - 0.00915(1.8T + 32) - \log[Ca2^{+}] - \log(\text{alkM}) + 0.1\log(\text{TDS})$	
Porosity of scaling	$p \cong \frac{[10 - (LSI)^2]}{10}$	
Uniform corrosion rate in scaling freshwater (μm/y)	$C_{\text{rate,water,LSI}>0} \cong 1.2 \cdot \{[O_2] + 0.04 \cdot [Cl_2]\} \cdot [10 - (LSI)^2] \cdot 2^{\frac{T-25}{25}}$	
Resistivity of freshwater (Ωm)	$\rho \cong \frac{9}{(1 + 0.02\Delta T) \cdot \text{TDS}}$	
Corrosion rate (mm/y) by differential aeration in freshwater	$C_{\text{rate,DA,FW}} \cong 9\sqrt{\frac{i_L}{\rho}}$ $C_{\text{rate,DA,FW}} \cong 0.1 \cdot \sqrt{[O_2] \cdot 2^{\frac{T-25}{25}} \cdot [10 - (LSI)^2] \cdot \text{TDS} \cdot (1 + 0.02\Delta T)}$	
Corrosion rate (mm/y) by galvanic corrosion in freshwater	$C_{\text{rate,GC}} \cong 20\sqrt{\frac{i_L \cdot \Delta V}{\rho}}$ $C_{\text{rate,GC,FW}} \cong 0.15 \cdot \sqrt{[O_2] \cdot 2^{\frac{T-25}{25}} \cdot [10 - (LSI)^2] \cdot \text{TDS} \cdot (1 + 0.02\Delta T)}$ $C_{\text{rate,GC,FW}} \cong 0.5 \cdot \sqrt{[O_2] \cdot 2^{\frac{T-25}{25}} \cdot (1 + \sqrt{v}) \cdot \text{TDS}} \cdot (1 + 0.02\Delta T)$	

(*Continued*)

Table 5.2 (Continued)

	Equation	Parameters
Corrosion rate in seawater (μm/y)	$C_{\text{rate,SW,stagnant}} \cong \frac{12 \cdot [O_2] \cdot 2^{\frac{T-25}{25}}}{\sqrt{t}}$ $C_{\text{rate,SW,flowing}} \cong 12 \cdot [O_2] \cdot (1+\sqrt{v}) \cdot 2^{\frac{T-25}{25}}$	$[O_2]$ (ppm) oxygen content v (m/s) water velocity T (°C) temperature t (y) exposure time for $t > 1$y
Galvanic corrosion rate in seawater (mm/y)	$C_{\text{rate,GC,SW}} \cong 1.4 \cdot \sqrt{[O_2] \cdot 2^{\frac{T-25}{25}} \cdot (1+\sqrt{v})}$	
Fluid critical velocity of water for erosion-corrosion occurrence (m/s)	$v_{E-C} \cong 3 \cdot k_M$ Cu 0.4 Cu–Ni (90/10) 0.9 Cu–Ni (70/30) 1.2 Mild steel (C–Mn steel) 1.2 Low alloy steel (Cr–Mo steel) 1.5 Monel 3 Stainless steel 4	v_{E-C} (m/s) critical velocity k_M metal constant
Corrosion rate (mm/y) by differential aeration in soil	$C_{\text{rate,DA,soil}} \cong \sqrt{\frac{[O_2]}{\rho}}$	$[O_2]$ (ppm) oxygen content ρ (Ωm) resistivity
Corrosion rate (mm/y) by galvanic corrosion in soil	$C_{\text{rate,GC,soil}} \cong 1.5\sqrt{\frac{[O_2]}{\rho}}$	
SRB corrosion rate (mm/y)	$C_{\text{rate,SRB}} \cong [SO_4^{2-}]^{0.5} - 0.0016 \cdot (T-38)^2$	$[SO_4]$ (ppm) sulphate content T (°C) temperature ($0 < T < 60$°C)

Corrosion in atmosphere Thickness loss at end of design life, *DL*	$\Delta s \cong \sum_{i=1}^{DL} WF_i \cdot C_{rate,i} \cdot \Delta t_i$	Δs (μm) total thickness loss WF_i wetting factor at time i $C_{rate,i}$ (μm/y) corrosion rate DL (y) design life Δt_i (y) exposure time $C_{rate,atm}$ (μm/y) corrosion rate $C_{0,atm}$ (μm/y) initial corrosion rate T (°C) temperature of surface of metal Cl (μg/m^3 in atmosphere) SO_2 (μg/m^3 in atmosphere) k_{atm} (μm/y) metal constant
Trend with time of corrosion rate in atmosphere (μm/y)	$C_{rate,atm} \cong \frac{C_{0,atm}}{\sqrt{t}}$	
Initial corrosion rate in atmosphere (μm/y)	$C_{0,atm} \cong k_{atm} \cdot [100 + (Cl + SO_2)] \cdot 2^{\frac{T-25}{25}}$ $k_{atm,Fe} = 1;$ $k_{atm,CorTen} = 0.5$ $k_{atm,Zn} = 0.04$ $k_{atm,Cu} = 0.02$	

5.8.2 Metals for freshwater

The materials used for the pipelines carrying freshwater are cast iron, reinforced concrete, carbon and low alloy steels, galvanized steel. The choice of material depends on either economic appraisal or technical factors, in particular diameter and pressure. For example, prestressed cylinder concrete pipes are used for large diameter, low pressure service pipes; cast iron for 20 bar max pressure tubes; for high operating pressures, steel and galvanized steel are used.

In industrial standards, mild steel, cast iron or galvanized steel are selected on the basis of parameters indicated in Table 5.A.1. Above 40°C galvanized steel may suffer galvanic corrosion because of so-called inversion of polarity, for which zinc becomes cathodic with respect to carbon steel, then enhancing galvanic attack on zinc coating defects.

5.8.3 Corrosion in water injection plants

In oil-&-gas plants, water injection is a common secondary enhancement technique, consisting of injecting treated water in the hydrocarbon reservoir. Most important requirements are: a low dispersed solid content to avoid reservoir plugging and, directly linked, a very low corrosiveness, since corrosion products increase dispersed solid content.

Water treatment, regardless its source (freshwater, seawater, brackish and formation water), includes: filtration (a coarse one by sand filter and a final by cartridge filter) oxygen removal and biocide injection. To reduce to a minimum the amount of corrosion product, a target of 10 μm/y as maximum accepted corrosion rate can be adopted which correspond to an amount of oxide of about 130 kg/y (or 15 mg/h) for each 4″ tubing, 5000 m deep. For this calculation, Eq. (1.14) is used ($1\ mg/m^2\ h \cong 1.1\ \mu m/y$).

To achieve this target, the maximum residual oxygen content is 0.3 ppm, given by Eq. (5.5) by inputting 10 μm/y as maximum corrosion rate, 4 m/s as maximum water velocity. A value about ten times lower, i.e. below 50 ppb, is considered in practice.

Table 5.A.1 Conditions for acceptable corrosion rate for mild steel and galvanized steel in freshwater [7,8]

Parameter—Index		Carbon steel	Galvanized steel
pH		>7	>7
Alk M	(mEq/L)	>2	>2
Acidity P	(mEq/L)	<0.05	<0.7
Ca	(mg/L)	>20	>20
O_2	(mg/L)	<0.02	<0.1
$HCO_3^-/(SO_4^{2-}+Cl^-)$	(mEq/L)	>1.5	>1
$(2SO_4^{2-}+Cl^-)/AlkM$	(mEq/L)	<5	<1

5.8.4 Corrosion in boilers

Corrosion control in boilers is carried out following three purposes: reduce conductivity by controlling salt content, maintain pH in the range 9–9.5 and reduce oxygen content of about 10 ppb (not lower than 5 ppb to favour magnetite formation and not above 50 ppb to avoid abnormal oxide growth).

To estimate the corrosion rate due to oxygen, Eq. (2.16) should be used to take into account accurately the influence of temperature on the diffusion coefficient and the influence of velocity through the Sherwood number. An acceptable estimation is also obtained by applying the simplified Eq. (2.11); therefore, by inputting an oxygen content of 10 ppb, a water velocity in tubes of 2–3 m/s and a temperature to give a factor of about 1000 (roughly 2^{10}), corrosion rate is 0.3 mm/y as maximum, which in practice reduces due to the presence of the magnetite film to about 50 μm/y.

To evaluate the incidence of acidic corrosion at pH close to 10, Eq. (2.4) can be used by inputting a temperature of about 250°C, a water velocity of 2–3 m/s and an iron ion concentration around 10^{-6} mol/L: a few μm/y is obtained, hence confirming that corrosion in boilers, once pH is under control, is a matter of residual oxygen content, only.

5.8.5 Case study—Corrosion in freshwater circuit

A circuit made of carbon steel, API Grade B, diameter ranging between 0.1 and 0.3 m, about 2000 m long, transports freshwater from supply wells to a discharge tank. After 4 y from commissioning, a few perforation occurred on the bottom of pipe, having a wall thickness of 4 mm. Operating was discontinuous. Cause of perforation was corrosion as ascertained by inspection.

Water analysis

pH: 8.5
Temperature: room temperature ($\approx$25°C)
Conductivity: 700 μS/cm (corresponding to TDS$\cong$0.65 g/L)
Alkalinity, alkM: 300 mg/L HCO_3
Chloride: 50 mg/L
Sulphate: 70 mg/L
Oxygen in water table: 0.2 mg/L
Oxygen in circuit: 0.4 mg/L
Redox potential: positive $\approx +6$ mV
SRB: absent
Langelier Index (calculated): positive$\approx$1.5.

Corrosion assessment

Corrosion attack was localized. Corrosion rate was about 1 mm/y. Possible root cause of localized corrosion is:

- Differential aeration.
- Galvanic corrosion assisted by oxides with noble behaviour as magnetite formed during welding.
- MIC due to SRB.

Generalized corrosion occurred all over the internal surface at a corrosion rate given by Eq. (5.7) for LSI > 0 giving an expected corrosion rate of 2 μm/y (LSI = 1.5; oxygen 0.2 ppm, room temperature).

Differential aeration

Corrosion rate is given by Eq. (5.11) giving a corrosion rate $C_{rate,DA}$ = 100 μmm/y.

Galvanic corrosion

Corrosion rate is given by Eq. (5.13a) giving a maximum corrosion rate $C_{rate,GC}$ = 180 μmm/y.

MIC

Conditions for MIC occurrence due to SRB are:

- anaerobic;
- redox potential negative;
- presence of nutrients and sulphates above 0.05–0.10 g/L.

Expected corrosion rate is zero for lack of active bacterial colonies; however, the maximum corrosion rate given by Eq. (5.22) would be $C_{rate,SRB} \cong 10$ μmm/y at 25°C and ≅250 μmm/y at 35–40°C.

Preliminary conclusion

Based on design data, experienced corrosion rate is not justified. Furthermore, by excluding MIC, localized corrosion by differential aeration or by galvanic coupling can give rise a corrosion rate of 1 mm/y if oxygen content would be much-much higher than concentration measured in water table.

By applying Eqs (5.11) and (5.13a), a corrosion rate of 1 mm/y is obtained if:

- for differential aeration, $[O_2]$ would be about 20 ppm;
- for galvanic corrosion, $[O_2]$ would be about 8 ppm.

As oxygen solubility in freshwater at room temperature is 10 ppm as maximum, galvanic corrosion could be occurred if oxygen entered the circuit and a galvanic coupling established.

Most likely cause of oxygen entry was partial water discharge by leaving the pipe partially empty. In fact, inside pipe, the water exposed to air can reached rapidly oxygen saturation and then maintained; moreover, galvanic effect arose for the presence of magnetite formed during welding.

Conclusion

Root cause of perforation was galvanic corrosion due to the presence of oxygen and a magnetite film; presence of oxygen derived from noncontinuous operating combined to a partially empty condition of piping.

5.8.6 Case study—Design of a loop circuit for corrosion testing

A loop circuit for corrosion testing was designed for the experimental evaluation of corrosion rate of metals used in industrial plants in water conveying piping. The loop was made of plastic with some inserted spools made of tested metals, specifically: mild steel, low alloy steel (2Cr–0.5Mo), copper, copper–nickel alloys

(90−10 and 70−30) and stainless steel, 316 grade. Aim of testing was to measure corrosion rate by changing the three influencing parameters: oxygen content, temperature and water velocity.

Corrosion conditions differ drastically whether or not erosion-corrosion condition sets up; this occurrence depends on water velocity in each spool: since the flow rate in the loop is fixed, velocity of water in each spool depends on spool diameter.

Flow rate could range from 1 to 5 L/s; hence, the diameter of each spool is obtained by calculating the spool cross section area from the maximum flow rate (5 L/s) and the critical erosion-corrosion velocity, in accordance with Eq. (5.18). Critical velocity for specified metals are as follows: mild steel, 3.6 m/s; low alloy steel (2Cr−0.5Mo), 4.5 m/s; copper, 1.2 m/s; copper−nickel alloys (90−10), 2.7 m/s and for (70−30), 3.6 m/s; 12 m/s for stainless steel, 316 grade.

Minimum spool diameter for each metal is the following:

- 42 mm, mild steel;
- 38 mm, low alloy steel (2Cr−0.5Mo);
- 73 mm, copper;
- 49 mm, copper−nickel alloys (90−10);
- 42 mm, copper−nickel alloys (70−30);
- 23 mm, stainless steel, 316 grade.

To study galvanic effects, two spools were jointed through an insulating flange so that galvanic current could be measured by an external electrical bonding through a calibrated shunt. For example, two couplings were compared *copper/mild steel* and *stainless steel/mild steel*. Aerated freshwater was tested, with resistivity of 20 Ωm. As expected, galvanic effect was almost the same for the two couplings at the maximum velocity for copper (about 1.5 m/s) for copper/mild steel couple and at the maximum velocity for mild steel (about 3.6 m/s) for the other couple, stainless steel/mild steel. Conversely, at very low velocity or near stagnant conditions the galvanic effect was much stronger for copper/mild steel couple, for about three times up to an order of magnitude. The obvious reason is the following: at high velocity (i.e. high turbulence) oxygen reduction reaction is under diffusion control on both metals, hence cathodic current density is about the same (i.e. oxygen limiting current density) while in stagnant condition, oxygen reduction is under activation control, so accordingly, cathodic current density is about 10 times lower on stainless steel than on copper, whilst driving voltage, ΔV, remains approximately constant (refer Fig. 1.6, by comparing copper against chromium). As a result, galvanic effect on carbon steel worsens when coupling metal turns from stainless steel to copper.

5.8.7 Case study—Corrosion assessment for atmospherically exposed tubular goods

Spare tubular goods are stored in a yard, freely exposed to the air. After 5 years, it was asked what would be the expected corrosion damage, as average and maximum thickness loss, if storage would be extended to 20 years.

Accordingly, it was decided to check thickness loss through an inspection and collect the average exposure conditions. The latter resulted as follows:

- chloride and sulphate content in the air (average sum was 355 ppm);
- average wetting factor, WF, within the year (33% in the year or percentage of time with RH higher than 85%);
- average temperature (18.4°C).

The inspection consisted in sampling and measurements as follows:

- number of specimens, for statistical random sampling on the basis of a confidence level, *a*, of 90% and a prevalence, *p*, of 10% is 22, given by Log*a*/Log*p* (refer to Eq. (8.27) of Chapter 8: Statistical Analysis of Corrosion Data);
- for each specimen, thickness loss was measured; mean and standard deviation resulted as 240 and 55 μm, respectively (CV is 0.23 which is close to 0.2 often adopted for atmospheric related corrosion);
- maximum value on stored spare parts after 5 years is 350 μm (given by $x_{max} \cong \mu + 2\sigma$).

Initial corrosion rate, C_0, was estimated by applying Eq. (5.25), which gives $C_0 = 380$ μm/y; the expected thickness loss after 5 year exposure is given by Eq. (5.23) which gives 280 μm as average and 390 μm as maximum on the basis of a standard CV = 0.2; this result is in good accordance with the measured values (240 μm as mean and 350 μm as maximum value).

As the results obtained by Eqs (5.24) and (5.23) were validated by inspection measurements, the forecast of expected thickness loss after 20 y appears reliable enough to count on. Accordingly, it was concluded that the calculated values of 560 μm as average and 780 μm as maximum were acceptable; nevertheless, it was recommended to apply a phosphatizing treatment every 5 y to half the expected thickness loss.

References

[1] Bianchi G, et al. Problemi di corrosione nei condensatori di vapore delle centrali termiche e nucleari, La Metallurgia Italiana, Ed. AIM, Atti e Notizie n. 10, 1973, Milan, Italy (in Italian).
[2] API RP14E. Recommended practice for design and installation of offshore production piping systems. Washington, DC: American Petroleum Institute; 1991.
[3] Von Wolzogen CAH, Van der Vlugt L. The graphitization, of cast iron as an electrochemical process in anaerobic soils. Water 1934;18:147–65.
[4] Dickinson WH, Caccavo Jr F, Lewandowski Z. The ennoblement of stainless steel by manganic oxide biofouling. Corros Sci 1996;38(8):1407–22.
[5] Linhardt P. Microbially influenced corrosion of stainless steel by manganese oxidizing microorganisms. Mater Corros — Werkstoffe und Korrosion 2004;55(3):158–63.
[6] ISO Std 9223, Corrosion of metals and alloys — Corrosivity of atmospheres — Classification, determination and estimation, Second Edition, 2012-02-01.
[7] Nace int. Publication, Corrosion and corrosion control in drinking water systems, Proc. Corrosion Workshop and Seminar, Oslo, Norway, March 19–21, 1990.
[8] EN 12502 – 2005, Protection of Metallic Materials against corrosion — Part 1,2,3,4,5, CEN/TC 262, Brussels.

Computer Modelling

6

Corrosion rate established by a macrocell mechanism was referred to in Chapter 3, Localized Corrosion by means a simplified empirical approach based on the estimation of the throwing power. The electrical field established by the macrocell is governed by the Laplace equation, whose analytical solution is possible for simplest geometries only. Nowadays, computer modelling is available, allowing an accurate estimation of the electrical field for all geometries then giving an accurate calculation of corrosion rates as well as current and potential distributions. Main applications deal with design and monitoring of cathodic protection (CP).

6.1 Electrochemical models

In the past, current and potential distributions were obtained by electrochemical models, which are nowadays replaced by mathematical methods on computer programs. However, it is worthwhile referring to them briefly for their historical importance in electrochemistry.

Primary distribution, which depends on geometry only, as discussed later on, can be easily obtained in laboratory through experimentations on a scale model, by taking into account that the relationships between parameters of the real system and those of the model are linear. Starting from the general relationship:

$$\Delta V = \rho \cdot i \cdot d \tag{6.1}$$

where symbols are known. By introducing the ratios of such parameters between real system, S, and model, M, relation (6.1) can be expressed as the following:

$$\begin{aligned} (\Delta V_S / \Delta V_M) &= (\rho_S / \rho_M) \cdot (i_S / i_M) \cdot (d_S / d_M) \\ r_V &= r_o \; r_\rho \; r_i \end{aligned} \tag{6.2}$$

where r is the ratio of parameters. Therefore, through this equation, it is possible to obtain current and potential distributions in any similar electrochemical system, once parameter ratios are known. For instance, by imposing the variation of conductivity ratio, $1/r_\rho$, proportionately with the size ratio, r_o, and by imposing same voltage to both systems (the real and the model), current densities and potentials are the same in corresponding points, so that it is sufficient to measure them on the model.

Fig. 6.1 illustrates current and equipotential lines obtained by Giuseppe Bianchi in a pioneering research work for a CP system for a buried pipeline through a model in 0.5 M lead sulphammic solution with lead electrodes, which behaves as pure ohmic system [1].

Engineering Tools for Corrosion. DOI: http://dx.doi.org/10.1016/B978-0-08-102424-9.00006-9

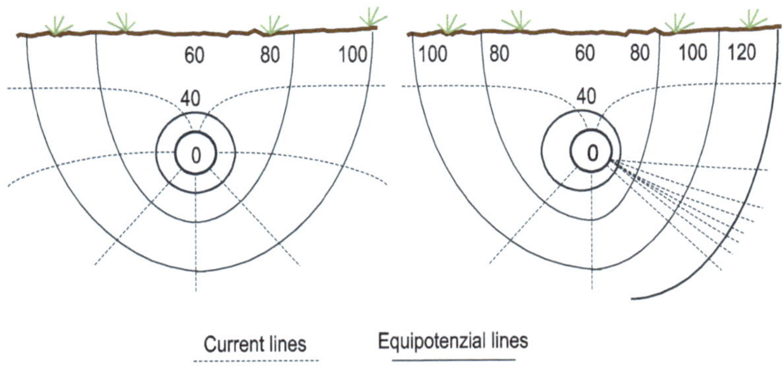

Figure 6.1 Current and potential distribution of CP of a pipeline obtained by electrochemical models: (A) with integral coating; (B) with a defect in the coating. (Values indicated are ohmic drops in mV.)
Adapted from Bianchi G., On the Stray Current in Soil and Current Distribution on Cathodic Protection, L'Elettrotecnica, 44, 20, p. 592.0; 1957. (in Italian).

The electrochemical model approach fails when overvoltage effects are no longer negligible with respect to the ohmic drop, because current distribution does not depend exclusively on system geometry. For all these cases, computer modelling has to be used.

6.2 Field equations

Electric field and all field-related problems are governed by the Poisson-Laplace "quasiharmonic equation" that, for stationary phenomena independent of time, assumes the general format:

$$\kappa \nabla^2 E + Q = 0 \tag{6.3}$$

where E is the potential function, Q is the electric charge flux, and κ is the conductivity taken constant. Besides the transportation of current as in the case of a macrocell, further examples of field problems are: heat transmission (E is temperature, Q is heat, κ is transport coefficient) and mass transport (E is pressure, Q is mass, κ is transport coefficient).

The derivative equation:

$$i = -\kappa \nabla E \tag{6.4}$$

links current density, i, and potential, E, so that equipotential lines and current lines are orthogonal to each other.

6.3 Macrocell

An electrochemical system as a macrocell is made up of an electrolyte and two or more electrodes between which the current is exchanged through electrode reactions.

The system can schematically be divided into two distinct domains: the *electrolytic solution* (the bulk) assumed to be a homogeneous ohmic conductor and the *double layer* at the electrode surface. By assuming the double layer negligible, the bulk is then governed by the *Laplace equation*, as an ohmic system:

$$\nabla^2 E_e = 0 \tag{6.5}$$

where E_e is the solution potential at the electrode–solution interface given by:

$$E_e = E_m - \eta \tag{6.6}$$

where E_m is the potential of a metal, which is uniform and constant (equipotential electrodes, assuming zero the ohmic drop in the metal), and η is the overvoltage which depends on current density, i, exchanged with the electrolyte through a general function as follows:

$$\eta = f(i) \tag{6.7}$$

The function $f(i)$ can be approximately linear as in the case of metal dissolution, especially for low-melting temperature metals, or logarithmic, in accordance with Tafel Law, as reported in Chapter 1, Basic Principles.

For the sake of simplicity, two situations can be envisaged although in electrochemistry a third condition is often referred to as tertiary distribution:

- *primary distribution* (as in the case of electrochemical models) for which overvoltage is negligible so that current distribution is governed by the ohmic drop in the electrolyte;
- *secondary distribution* for which overvoltage applies ($\eta \neq 0$).

6.3.1 Primary distribution (or ohmic systems)

In a purely ohmic conductor with resistivity ρ ($\rho = 1/\kappa$), the term Q is nil and the field equation is simplified to Laplace's equation:

$$\nabla(\kappa \nabla E) = 0; \quad \text{or} \quad \nabla^2 E = 0 \tag{6.8}$$

which gives Ohm's Law:

$$i = -\kappa \nabla E = -\frac{1}{\rho} \nabla E \cong \frac{\Delta V}{\rho} \tag{6.9}$$

Boundary conditions are reduced to $\partial E / \partial n = 0$, which means that the insulated surface does not exchange current.

Analytical solution of Laplace equation (6.8) exists for simple geometries, only, as for example parallel plates, concentric spheres, or coaxial cylinders. Table 6.1

Table 6.1 Primary current distribution for simple geometries [2]

		Potential	Current	Resistance
PARALLEL PLATES (large surface area) d = distance between plates S = surface area of plates	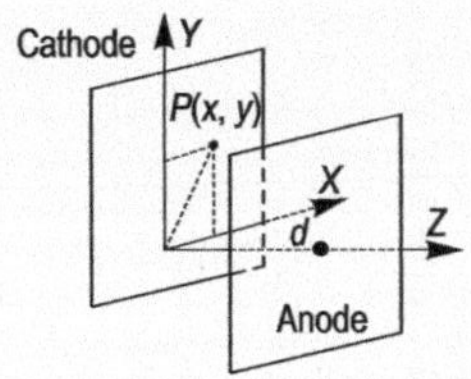	$E_{(z)} = \rho\, I \frac{Z}{S}$ $E_{(x,y)} = \text{const}$	Uniform $i_a = i_c = \frac{I}{S}$	$R = \rho \frac{l}{S}$
SPHERIC (Concentric Spheres) r_a; r_c = radius of spheres; r = distance from centre	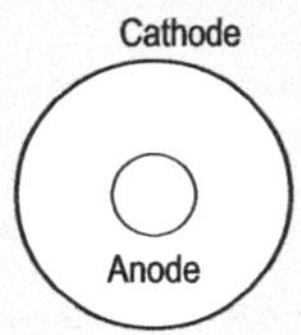	$E_{(r)} = \frac{\rho\, I}{4\pi}\left(\frac{1}{r_a} - \frac{1}{r}\right)$	Uniform $i_a = \frac{I}{4\pi\, r_a^2}$; $i_c = \frac{I}{4\pi\, r_c^2}$	$R = \frac{\rho}{4\pi}\left(\frac{1}{r_a} - \frac{1}{r_c}\right)$ $R \cong \frac{\rho}{4\pi\, r_a}$; (if $r_a \ll r_c$)
CYLINDRICAL (coaxial cylinders) $L \gg r_c$)	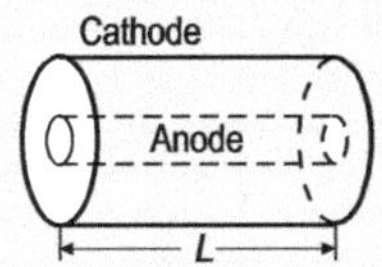	$E_{(r)} = \frac{\rho I}{2\pi\, L} \ln \frac{r}{r_a}$	Uniform $i_a = \frac{I}{2\pi\, L\, r_a}$; $i_c = \frac{I}{2\pi\, L\, r_c}$	$R = \frac{\rho}{2\pi\, L} \ln \frac{r_c}{r_a}$
INFINITE PLATE (cathode) Small anode at distance d on Z axes r = distance of point $P(x, y)$ from anode projection	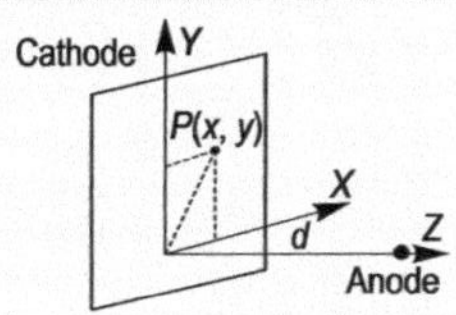	$E_{(z)} = \frac{\rho I}{2\pi}\left[\frac{z}{d^2 - z^2}\right]$	NONuniform $i_{\max} = \frac{I}{2\pi\, d^2}$; $i = i_{\max}\left(\frac{d}{\sqrt{d^2 + r^2}}\right)^3$	$R \cong \frac{2\rho}{\pi\, d}$

INFINITE PLATE (cathode) Linear anode at distance d on Z axes and infinite length r = distance of point $P(x, y)$ from anode projection	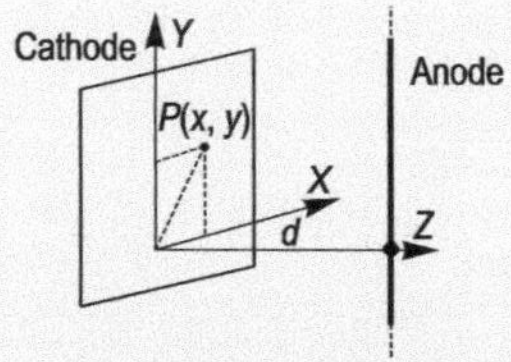	$E_{(z)} = \frac{\rho}{\pi}\left(\frac{I}{L}\right)\ln\frac{d+z}{d-z}$ $E_{(y,x)} = \frac{\rho}{2\pi}\left(\frac{I}{L}\right)\ln\frac{(d+z)^2+x^2}{(d-z)^2+x^2}$	NON uniform $i_{\max} = \frac{1}{\pi d}\left(\frac{I}{L}\right);$ $i_{(x)} \cong i_{\max}\frac{d^2}{d^2+x^2}$	$R_{(L)} \cong \frac{\rho}{\pi L}$ (if $d \gg$ anode diameter)
FLAT DISK (cathode radius r_o) Anode at infinite distance d = distance from disk	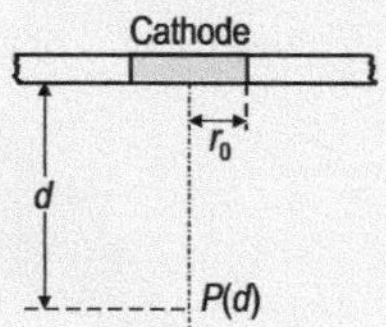	$E_{(d)} \cong \frac{\rho I}{4\pi d}$ with $r_o \ll d$	Approximately uniform $i \cong \frac{I}{2\pi r_o^2}\sqrt{\frac{1-r_o^2}{r_o^2}}$	$R \cong \frac{\rho}{4\pi r_o}$

reports Laplace equation solutions for such simple geometries which give the primary current distribution [2–5]. Important features of primary current are as follows:

- Its distribution depends only on geometric factors and not on environmental resistivity;
- Ohmic drop is concentrated in a volume close to electrodes. As a rule of thumb, this effect prevails on smaller electrode.

To explain the effects of primary current distribution, reference is made to Fig. 6.2, where a macrocell is represented (e.g. a CP system). The balance of voltage for each current path is the following:

$$E^{AC} = E^{AC}_{eq} + \Psi_a + |\Psi_c| + \Psi_{ohm} \tag{6.10}$$

where ψ_a, ψ_c, are anodic and cathodic overvoltage (calculated from the related free corrosion potentials) and ψ_{ohm} is ohmic dissipation, i.e. ohmic drop. If primary distribution applies (both cathodic and anodic overvoltage are negligible) and electrolyte is homogeneous, Eq. (6.10) becomes:

$$(E^{AC}_{eq})_1 + \Psi_{ohm,1} = (E^{AC}_{eq})_2 + \Psi_{ohm,2} \tag{6.11}$$

which establishes the condition of constant ohmic drop along each current path:

$$\Psi_{ohm,1} = \Psi_{ohm,2} \tag{6.12}$$

This ohmic drop is expressed by the Ohm Law as follows:

$$\Psi_{ohm,n} = I_n\, R_n = \rho \int_n \frac{i_n}{S}\, dL \tag{6.13}$$

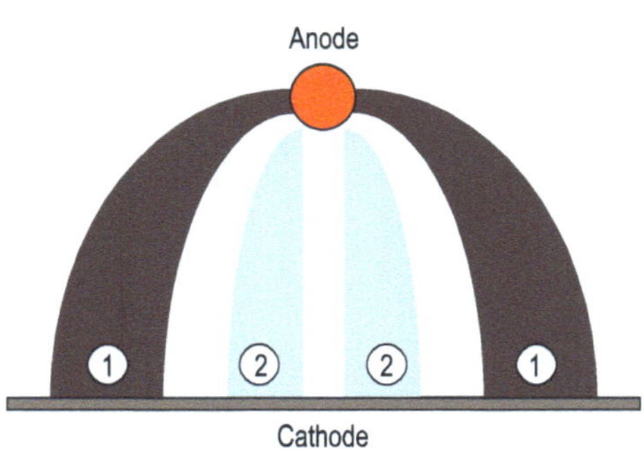

Figure 6.2 Scheme of current distribution in a macrocell system.
From Lazzari L., Pedeferri P., Cathodic Protection. Polipress. Italy: Milan; 2005. ISBN 88-7398-020-1.

where I_n is the circulating current in flux tube nth of length L and section S over the cathodic surface; therefore, $I_n = i_n S$ varies with L. As ohmic drop is constant, from Eq. (6.12), the current density ratio for different current paths is obtained:

$$\frac{i_1}{i_2} \cong \frac{\int_2 \frac{dL}{S}}{\int_1 \frac{dL}{S}} \tag{6.14}$$

Eq. (6.14) shows that an increase in flux tube length leads to a decrease in current density and therefore the system is characterized by a poor throwing power. A practical uniform primary distribution sets up when the term $\int dL/S$ is almost constant for all current paths.

6.3.2 Localization and calculation of the ohmic drop

For ohmic systems, i.e. overvoltage is discarded, it could be useful to refer to an approximate calculation of the ohmic drop as given by Eq. (6.13) when the following conditions apply:

- The domain is large; hence, ohmic drop is concentrated at the electrodes. For example, in the case of a cathodically protected pipeline by a remote anode, ohmic drop concentrates near both electrodes because the potential gradient zeros in the between space, as integral $\int dL/S$ zeros when section area tends to infinite;
- If one of the electrode is small (usually the anode), the total resistance practically coincides with the one associated to that.

The ohmic resistance is expressed, with an error lower than about 10%, by empirical equations which are function of environment resistivity and size of the smaller electrode (typically the anode) by the following general expression:

$$R_a = K\,\rho f(\Gamma) \tag{6.15}$$

where Γ is characteristic electrode size (in the case of a rod type anode, its length and in the case of a disk shape anode, its diameter) and ρ is resistivity of electrolyte surrounding the electrode. Table 6.2 lists the empirical formulae used to calculate anodic resistance in galvanic and impressed current CP systems [2,6].

6.3.3 Secondary distribution

When electrode polarization is established, solution of the Laplace equation has to take into account the boundary conditions between overvoltage and current density. In general terms, secondary distribution is a function of the following type [2,3]:

$$\frac{i}{i_{av}} = f\left(\Gamma, \rho, \frac{d\eta}{di}\right) \tag{6.16}$$

Table 6.2 Summary of equations used for calculations of anode resistance and potential trend [2,6]

Anode shape		Resistance	Potential
SEMISPHERICAL Radius r_o	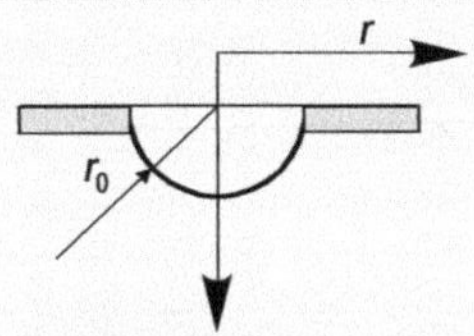 	$R=\frac{E_o}{I}=\frac{\rho}{2\pi r}$	$E_r=E_o\frac{r_o}{r};\ (r>r_o)$
SPHERICAL Diameter ϕ Burial depth H	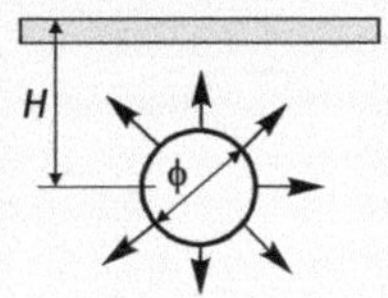 	$R=\frac{\rho}{2\pi}\left(\frac{1}{\phi}+\frac{1}{4H}\right)$ $R=\frac{\rho}{2\pi\phi};\quad (\text{if } H\gg\phi)$	$E_r=E_o\frac{\rho}{2\pi R}\sqrt{H^2+r^2}$
FLAT CIRCULAR Radius r_o	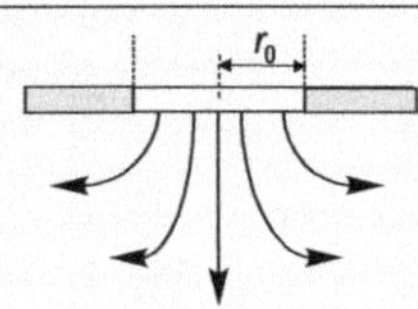 	$R=\frac{E_o}{I}=\frac{\rho}{4\pi r_o}$	$E_r=2\pi E_o \text{ arcsen}\left(\frac{r_o}{r}\right)$ $E_x=E_o\frac{r_o}{x};\quad (\text{if } x\gg r_o)$
FLAT OVERLAPPED Radius r_o or sizes a and b	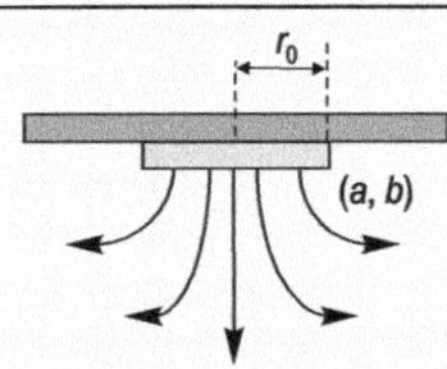 	$R=\frac{E_o}{I}\cong\frac{\rho}{4r_o}$ $R\cong\frac{\rho}{(a+b)}$	$E_r=2\pi E_o \text{ arcsen}\left(\frac{r_o}{r}\right)$ $E_x=E_o\frac{r_o}{x};\quad (\text{if } x\gg r_o)$

FLAT SUSPENDED Radius r_o or sizes a and b	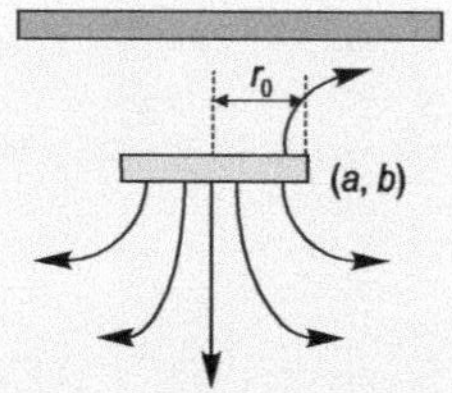	$R = \frac{E_o}{I} \cong \frac{\rho}{8r_o}$ or $R \cong \frac{\rho}{2(a+b)}$	$E_r = 2\pi E_o \text{ arcsen}\left(\frac{r_o}{r}\right)$ $E_x \cong E_o \frac{r_o}{x}$; (if $x \gg r_o$)
VERTICAL SUPERFICIAL Length L Diameter ϕ	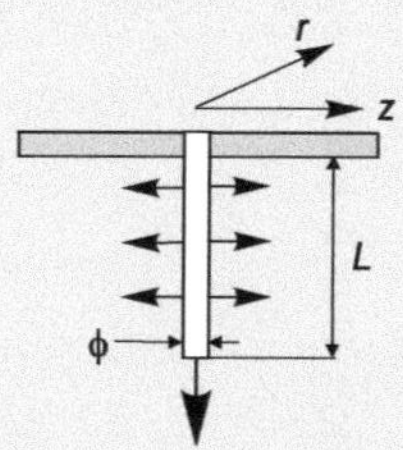	$R = \frac{\rho}{4\pi L}\left(\ln\frac{4L}{\phi}\right)$ $(r > L \gg \phi)$	$E_r = E_o \frac{L}{r\left(\ln\frac{4L}{\phi}\right)}$ $E_x = E_o \frac{\rho}{4\pi LR}\left(\ln\frac{x+L}{x-L}\right)$
HORIZONTAL SUPERFICIAL Length L Diameter ϕ	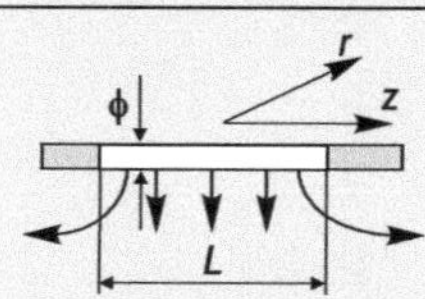	$R = \frac{\rho}{4\pi L}\left(\ln\frac{2L}{\phi}\right)$ $(r > L \gg \phi)$	$E_r = E_o \frac{L}{r\left(\ln\frac{2L}{\phi}\right)}$ $E_x = E_o \frac{\rho}{4\pi LR}\left(\ln\frac{x+L}{x-L}\right)$
DEEP VERTICAL OR HORIZONTAL Length L Diameter ϕ Burial depth H	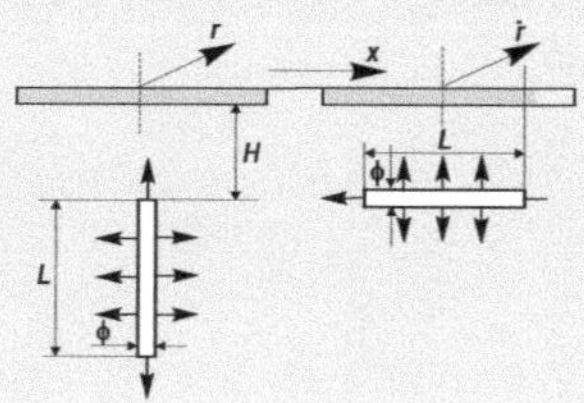	(Dwight Equation) $R = \frac{\rho}{2\pi L}\left(\ln\frac{8L}{\phi} - 1\right)$ $(L \gg \phi;\ H \gg L)$	$E_r = E_o \frac{L}{r\left(\ln\frac{8L}{\phi}\right)}$ $E_x = E_o \frac{\rho}{4\pi LR}\left(\ln\frac{x+L}{x-L}\right)$

where Γ is geometric factor, $d\eta/di$ is polarization (i.e. overvoltage as function of current density) and ρ is the electrolyte resistivity. As a general conclusion, secondary distribution is more uniform than primary, because geometric factors determining the primary distribution are overlapped by overvoltage effects that smooth the current distribution trend.

When the cathodic reaction is hydrogen evolution (in acid solutions or in overprotection conditions for CP systems), overvoltage is low (because Tafel slope is about 120 mV/decade) so that current distribution is closer to primary one; conversely, when the cathodic reaction is oxygen reduction, overvoltage is much higher, on the order of one volt, so that secondary distribution prevails. This typically occurs in CP in natural environments, especially in low resistivity solutions such as seawater, where overvoltage largely exceeds ohmic drop contribution.

6.3.4 Current density close to an electrode (anode)

Before introducing the computer modelling and relevant results, it is useful to outline the empirical procedure for calculation of the corrosion rate, as anodic current density in close proximity to the anode, from the primary current distribution. From Eq. (6.9), in close proximity to the anode, Laplace equation shortens to:

$$i_a \cong \frac{\Delta V}{\rho \cdot d} \tag{6.17}$$

where current density, i_a (mA/m^2), is anodic current density close to the anode; ΔV (mV) is ohmic drop at distance d (m) from the anode and ρ (Ωm) is electrolyte resistivity. Eq. (6.17) fails when distance, d, is much higher than anode size.

A correction of Eq. (6.17) when high cathode-to-anode surface area ratio applies, can deal with a cylindrical current distribution which leads to the following relationship (see Table 6.1):

$$E(d) = \frac{\rho}{2\pi} \cdot \frac{I}{L} \cdot \ln\frac{d}{r_a} \tag{6.18}$$

where $E(d)$ (mV) is the potential at a specified distance, r_a (m) is the anode size (radius); I/L (mA/m) is specific anode output, ρ (Ωm) is the electrolyte resistivity and d is the distance from anode. If ΔV is ohmic drop between two points, spaced d (m), in radial direction, anodic current density (i.e. corrosion rate) is given from Eq. (6.18) as follows:

$$i_a \cong \frac{\Delta V}{\rho \cdot r_a \cdot \ln\frac{r_1}{r_2}} \tag{6.19}$$

where ΔV (mV) is the ohmic drop; i_a (mA/m^2) is the anodic current density close to the anode; r_a (m) is the anode size (radius) and $r_1 - r_2 = d$ is reference electrode spacing ($r_2 > r_a$).

6.4 Boundary conditions

To solve the Laplace equation, numerical methods as finite element method (FEM), finite difference method (FDM) and boundary element method (BEM) are used; the latter is often preferred because it only requires the input of sizes and current potential functions of anode and cathode, regardless the description of the electrolyte domain.

The key point for applying these methods is how to input electrochemical boundary conditions, particularly those related to CP in seawater, where calcareous deposit forms as polarization proceeds.

For seawater applications, the following boundary conditions are used [7]:

- galvanic anodes: constant potential condition (i.e. negligible anode polarization);
- impressed current anodes: constant current output condition;
- cathode: the boundary condition is the relationship $i = f(\eta)$ as illustrated in Fig. 6.3 and given by the following general expression derived from Butler–Volmer equation (refer to Chapter 1: Basic Principles):

$$i = i_{0,\mathrm{Fe}} \cdot e^{\frac{2.3 \cdot (E_{\mathrm{Fe}} - E)}{b_\mathrm{a}}} - i_\mathrm{L} - i_{0,\mathrm{H}_2} \cdot e^{\frac{-2.3 \cdot (E - E_{\mathrm{H}_2})}{b_{\mathrm{H}_2}}} \tag{6.20}$$

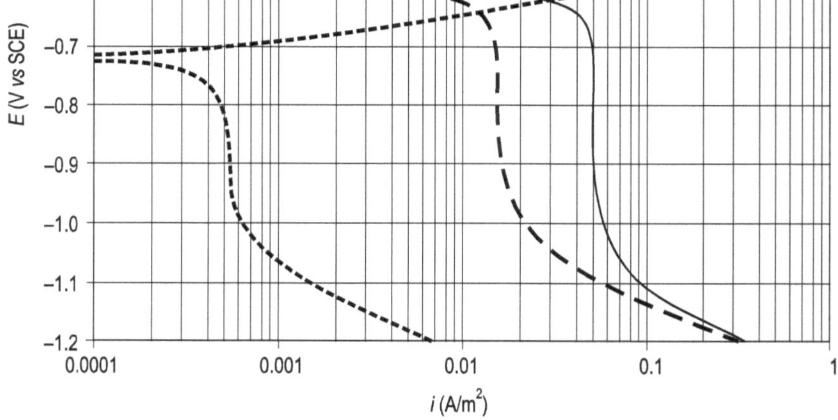

Figure 6.3 Plot of Butler–Volmer equation: (A) in flowing seawater ($i_L = 50$ mA/m^2); (B) in mud ($i_L = 15$ mA/m^2); (C) in the presence of a coating ($i_L = 0.5$ mA/m^2); $i_{o,\mathrm{Fe}} = 1$ mA/m^2; $i_{o,\mathrm{H}} = 1$ mA/m^2; b_{H_2}; $= 0.12$ V; $b_{\mathrm{Fe}} = 0.06$ V.

where $i_{0,Fe}$ (mA/m^2) is the iron exchange current density; $i_{0,H}$ (mA/m^2) is the hydrogen exchange current density, i_L (mA/m^2) is the oxygen limiting diffusion current density, b_{H_2} (V/decade) is the cathodic Tafel slope for hydrogen and b_{Fe} (V/decade) is the anodic Tafel slope for iron; E_{Fe} (V) is the equilibrium potential of iron, often approximated to corrosion potential and E_{H_2} (V) is equilibrium potential of hydrogen evolution reaction.

6.4.1 Time-dependent boundary conditions

Boundary conditions at the cathode change as polarization takes place due to calcareous deposit formation. In order to address this problem, it is useful to separate the two conditions:

- Stationary conditions: independent from time and in the absence of calcareous deposit; Eq. (6.20) is used for boundary conditions;
- Nonstationary conditions: when the calcareous deposit forms, following a time-dependent law for oxygen limiting current density, i_L. Laplace equation is still applicable assuming that boundary conditions change slowly through "quasistationary" steps, by which parameters are slightly influenced by the calcareous deposit and, therefore, may be considered constant.

The decay function of current density can be simply expressed by a relationship which links, by increasing steps, time and changes of current density as follows:

$$i_x(t + dt) = i_x(t) \cdot (1 + R_x dt) \tag{6.21}$$

where i_x is the current density at time t and R_x is the law of current density decrease over time. In initial phases, current density is very high due to the initial availability of oxygen, as it is already present in the diffusion layer. Once this oxygen is consumed, the oxygen flows slowly by diffusion and the current density decreases until it reaches the limiting current density. The function R_x also takes into account if calcareous deposit forms. To give an example for seawater applications, initial current density starts from very high value as 1 A/m^2 for a short time then decreasing within some months to about 10 mA/m^2, only.

6.5 FEM and BEM methods

Finite element method (FEM) and boundary element method (BEM) are strong numerical analysis tools for engineers for obtaining numerical solutions of field problems for complex geometries.

In a few words, the two methods diverge on where discretization is applied: in FEM it involves the domain, i.e. the electrolyte, whereas in BEM the surfaces of the sources, only, i.e. anode and cathode, regardless the domain. Because of this, BEM appears to be more appropriate, besides more elegant, for CP applications, in which anode and cathode surfaces are known.

In FEM, the domain, i.e. the electrolyte for example seawater, is subdivided into small elements to which Ohm Law is sequentially applied from the anode to the cathode. Once input the boundary condition on anode surface (typically constant potential and uniform current density) and those for cathode surface (Butler–Volmer equation or linearized condition) the potential along a current path is obtained as sum of ohmic drop of each element, ith, ($IR_i = i \cdot l \cdot \rho$, where ρ is the electrolyte resistivity, i is the current density through the element and l is the size of the element).

Instead, BEM requires boundary conditions on anode and cathode surfaces as for FEM, whereas domain is a sufficient large volume which includes the galvanic or macrocell system.

In both applications, the output is current density and potential distribution on electrodes, typically for CP on the cathodic surface.

6.6 Summary

Table 6.3 summarizes equations for the estimation of current and potential distribution and boundary conditions for Laplace equation.

6.7 Appendix

6.7.1 Case studies

Computer modelling has been used in a variety of CP applications. In the following, two case studies are presented for which the commercial software Comsol Multiphysics 5.2 was used.

6.7.1.1 Anode distribution on a jacket of an offshore platform

In CP applications for offshore structures as a jacket of a platform, classical empirical approach for anode distribution is simply based on experience following some practical rules of thumb, like:

- potential distribution is taken as uniform on the entire surface of the structure; hence, for calculation, the anode driving voltage is taken as constant;
- anode mass and maximum anode output are calculated in regards to the local pertinent surface area (e.g. for a brace the surface area of anode spacing) and constant protection current density, distinguishing between an initial value for achieving the polarization and a stationary value for calculating the anode consumption;
- empirically, anodes are located close to nodes; on braces and legs maximum spacing is 8–10 m depending on sea.

In the past until the 1970s, unsuccessful applications were experienced because protection current density was underestimated and no idea of the potential

Table 6.3 Summary of equations related to electric field and boundary conditions

	Equation	Parameters
Electrochemical model equation	$r_E = r_o\, r_\rho\, r_i$	r_o ratio size r_i ratio current density r_E ratio potential r_ρ ratio resistivity
Laplace equation	$\nabla^2 E_e = 0$ $E_e = E_m - \eta$	E_e (V) solution potential E_m (V) potential of a metal η (V) overvoltage i (A/m^2) current density κ (S/m) conductivity ρ (Ωm) resistivity
Ohmic systems	$i = -\kappa\nabla E = -\frac{1}{\rho}\nabla E \cong \frac{\Delta V}{\rho}$	
Ohmic drop along a current path	$\Psi_{ohm,n} = I_n R_n = \rho\int_n \frac{i_n}{S}\, dL$	nth flux tube I_n (A) circulating current L (m) length S (m^2) cross section area ρ (Ωm) resistivity Γ (m) geometrical size K constant
Ohmic resistance, R_a	$R_a = K\,\rho\, f(\Gamma)$	
Current density close to the electrode (anode)	$i_a \cong \frac{\Delta V}{\rho\cdot d}$	ΔV (mV) ohmic drop d (m) distance from electrode E (V) potential d (m) distance from electrode I/L (A/m) specific output r_a (m) electrode size ρ (Ωm) resistivity $r_1 - r_2 = d$ (m) electrode spacing; $r_2 > r_a$
	$E(d) = \frac{\rho}{2\pi}\cdot\frac{I}{L}\cdot\ln\frac{d}{r_a}$	
Current density close to the electrode (anode)	$i_a \cong \frac{\Delta V}{\rho\cdot r_a\cdot\ln\frac{r_1}{r_2}}$	
Butler–Volmer equation	$i = i_{0,Fe}\cdot e^{\frac{2.3\cdot(E_{Fe}-E)}{b_a}}$ $- i_L - i_{0,H_2}\cdot e^{\frac{-2.3\cdot(E-E_{H_2})}{b_{H_2}}}$	$i_{0,Fe}$ (mA/m^2) exchange c.d. $i_{0,H}$ (mA/m^2) exchange c.d. i_L (mA/m^2) limiting c.d. b_{H_2} (V/decade) Tafel slope b_{Fe} (V/decade) Tafel slope E_{Fe} (V) equilibrium potential E_{H_2} (V) equilibrium potential i_x (mA/m^2) c.d. at time t t (day) time of polarization R_x (1/day) decrease rate of c.d.
Nonstationary polarization conditions	$i_x(t+dt) = i_x(t)\cdot(1 + R_x dt)$	

Figure 6.A1 Example of potential distribution obtained by BEM computer modelling for a portion of a jacket of a platform $i_L = 150$ mA/m^2; $i_{o,Fe} = 1$ mA/m^2; $i_{o,H} = 1$ mA/m^2; b_{H_2}; = 0.12 V; $b_{Fe} = 0.06$ V.
Courtesy Cescor srl.

distribution was accessible; even the concept of the throwing power of the CP, i.e. of a macrocell, was not clearly established and then not employed. It appears evident that by following this empirical approach no information is available on the real potential distribution, especially on nodes where ohmic drop increases because of the narrower geometry.

A reliable anode distribution is obtained by the use of computer modelling, as Fig. 6.A1 shows, even for complex geometry as nodes. It is easy to foresee protection condition either at startup when protection current density is high (see Table 1.A2) or after achieving polarization through the decay function of Eq. (6.21). Parameters used for the Butler–Volmer equation (6.20) for case study of Fig. 6.A1 are: $i_L = 150$ mA/m^2; $i_{o,Fe} = 1$ mA/m^2; $i_{o,H} = 1$ mA/m^2; b_{H_2}; = 0.12 V; $b_{Fe} = 0.06$ V/decade. The simulation allows easily the change of anode position aiming to obtain a more uniform potential distribution.

6.7.1.2 Coating defect size estimation on a pipeline

Computer modelling allows to combine on-site potential measurements, well known as close interval potential profile (CIPP), and the estimation of the size of coating defects. In the following, a case study on a subsea pipeline survey is presented [8,9].

Submerged pipelines are protected by a combination of an external coating and bracelet-type galvanic anodes, having a spacing which is fixed by experience in the range 80 to 300 m; this spacing can be verified by estimating the throwing power of the galvanic cell through the empirical equations summarized in Table 3.1. Affecting parameters are the electrolyte resistivity (in seawater favourably low) and protection current density based on expected coating efficiency.

It appears evident that any coating defect can jeopardize the throwing power with the consequence of a risk of insufficient protection on the defect if its size exceeds a threshold value. It is common practice to verify periodically the protection potential all along the pipeline by means of the achievement of both the potential and current density profiles. These are obtained by the handling of ohmic drop measurements between three reference electrodes: one remote and the other two close to the pipeline by means of a probe shown in Fig. 6.A2, which measures the between ohmic drop. Fig. 6.A3 shows an example of potential measurement obtained by a CIPP survey provided an electrical connection with the pipeline (so-called trailing wire technique). Fig. 6.A4 shows an example of the ohmic drop profile obtained on field during a survey, where sea depth was about 16 m and the pipeline was laid on sea bottom. Negative peaks correspond to a coating defect (in

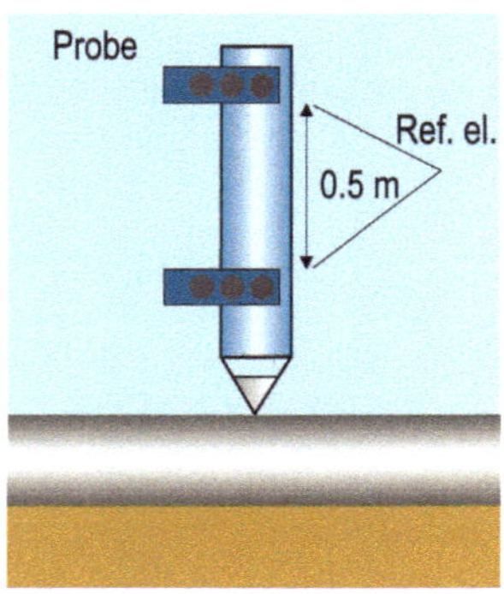

Figure 6.A2 Device for the measurement of ohmic drop profile.
Courtesy Cescor srl.

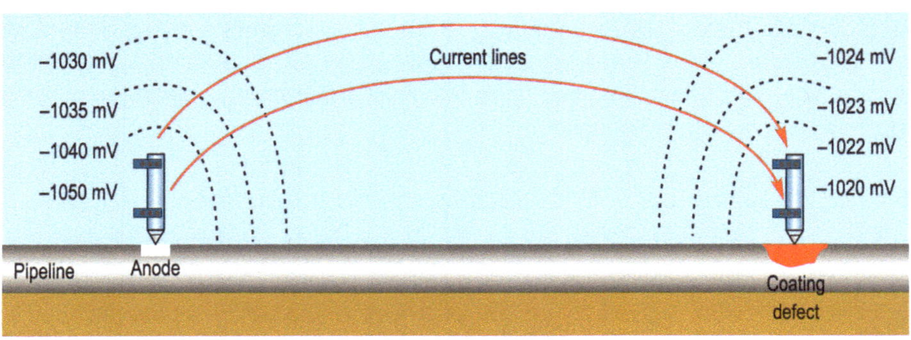

Figure 6.A3 Artistic reproduction of the CIPP survey pipeline [8].

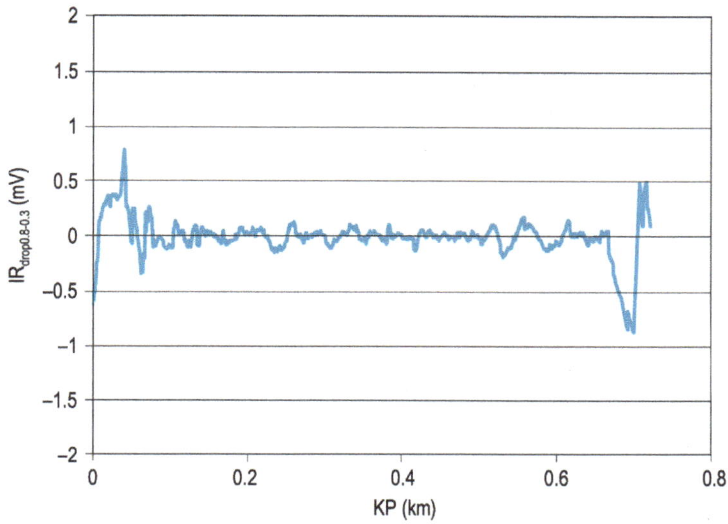

Figure 6.A4 Example of ohmic drop profile recorded on a submerged pipeline [8].

the plot at KP 0.7 km, with −0.8 mV ohmic drop), whereas positive peaks indicate anodes (in the plot at KP 0 km).

An estimation of the current density, either anodic or cathodic, can be worked out through the simplified Eq. (6.17) or by the improved one (6.19) for anodes. The potential profile is reconstructed from the anode potential to which the ohmic drop profile is added, assuming constant the remote potential. As far as the defect size is concerned, neither information nor calculation can be dealt with.

Conversely, by overlapping ohmic drop profile achieved by field surveys and FEM, a relationship between measured *IR* drop and the defect size can be obtained. In this case study, parameters used for the Butler−Volmer equation (6.20) are: $i_L = 15$ mA/m^2 (pipeline buried in mud); $i_{o,Fe} = 1$ mA/m^2; $i_{o,H} = 1$ mA/m^2; b_{H_2}; = 0.12 V; $b_{Fe} = 0.06$ V. FEM analysis results showed that an IR_{drop} of −0.8 mV indicates a defect size of about 152 mm in equivalent diameter. Indeed, this size was compatible with an uncoated flange on a 6″ pipeline which was found later on.

References

[1] Bianchi G. On the stray current in soil and current distribution on cathodic protection. L'Elettrotecnica 1957;44(20):592.0. (in Italian).

[2] Lazzari L, Pedeferri P. Cathodic protection. Italy: Milan: Polipress; 2005. ISBN 88-7398-020-1.

[3] Newman J. Mass transport and potential distribution in the geometry of localised corrosion, NACE-3. Houston: Texas; 1974. p. 45−61.

[4] Wagner C. Contribution to the theory of cathodic protection. J Electr Soc 1952;99(1):1.

[5] Kasper C. The theory of the potential and the technical practice of electrodeposition. Trans Electr Soc January 22, 1940;77:365–84. p. 1. Manuscript received.
[6] Beckman W, von Schwenk W. Handbook of cathodic protection. Redhill Surrey: Portcullis Press Ltd.; 1975ISBN 090199457X.
[7] Pastore T. Potential and current distribution of cathodic protection of offshore structures. Definition of boundary conditions, PhD thesis. Italy: Milan: Polytechnic of Milan; 1989. (in Italian).
[8] Marcassoli P, Bonetti A, Lazzari L, Ormellese M. Modelling of potential distribution of subsea pipeline under cathodic protection by finite element method. Mater Corros 2015;66(7). 2015 http://dx.doi.org/10.1002/maco. 201407738; 2015.
[9] Marcassoli P., Bazzoni B., Woodland P., Interpretation of galvanic anode inspection data through Finite Element Method (FEM) Modeling. Proceedings of Eurocorr2016 Conference. France: Montpellier; Sept 2016.

Testing

7

7.1 Classification of corrosion tests

Corrosion tests can be classified as follows:

- basic laboratory tests;
- quality control tests;
- pass–fail tests;
- accelerated tests.

7.1.1 *Basic laboratory tests*

The aim of these tests is the determination of critical values of basic corrosion-influencing parameters as those considered in journals and handbooks, as for example, thermodynamic and kinetic parameters, as anodic and cathodic characteristics, corrosion potential in referred electrolytes, tendency to passivation, oxidation power. As test conditions are rigorously controlled, results show a high reliability.

7.1.2 *Quality control tests*

They consist of standard tests required and carried out to check compliance with acceptance criteria. Like for standard mechanical testing, test conditions can differ from operating ones. Test conditions can be in accordance with norms, for example ISO, EN, ASTM or company standards. Among a variety of tests, most common known are, for example: Salt spray and acetic salt spray test for coating and painting acceptance, Huey test, Strauss test and Streicher test for intergranular corrosion susceptibility of stainless steels and nickel alloys.

7.1.3 *Pass–fail tests*

Many localized corrosion attacks, as for example stress corrosion cracking (SCC), crevice and pitting, follow a behaviour described by a *pdf* of log N type or very close to it. This implies that the occurrence of the attack happens during the first period of exposure (also known as *infant mortality*) rather than after long exposure time, as typically for ageing. In other words, should the attack take place, this does occur very soon or never.

Based on this behaviour, so-called pass–fail tests are tailored for a predetermined fixed exposure time: the criterion for interpretation is the following: test passed if within the exposure testing time no failures, or damage below a threshold, occurred.

A typical example is the SSC (sulphide stress cracking) test in compliance with normative [1], which specifies a testing time of 720 h and if neither failure nor secondary

Engineering Tools for Corrosion. DOI: http://dx.doi.org/10.1016/B978-0-08-102424-9.00007-0

cracks would appear, the test is considered passed, although theoretically a failure might occur in longer exposure time. Moreover, hydrogen induced cracking (HIC) test [2] follows a similar approach: test passed if no stepwise cracking or cracking below an agreed threshold after a fixed exposure time of 96 h occurred.

7.1.4 Accelerated tests

To validate the metal selection carried out in design, exposure tests are often required. They consist of testing candidate metals in a testing fluid, through two procedures: same process condition or so-called accelerated conditions.

The principle of an accelerated test is summarized in Fig. 7.1, where "*S*" and "*R*" are applied and resistance values, respectively, for operating and testing. In normal operating, no failure is expected, instead by increasing an influencing parameter failure occurs because resistance is exceeded in overlapping zone.

The accelerating corrosion-related parameter and testing conditions should be conveniently selected by complying with following general conditions:

- corrosion mechanism is the same (for instance, same cathodic process);
- there is a relationship between the measured parameter (i.e. time) and the increased parameters (i.e. temperature, acidity, turbulence and oxygen content);
- probability density distribution of results is the same (i.e. normal or log *N* and others).

Last condition derives from reliability and is clearly represented in Fig. 7.2 when extreme value statistics (generally Weibull or Gumbel) is adopted (see Chapter 8: Statistical Analysis of Corrosion Data).

As general approach for representing and elaborating testing data, an Arrhenius relationship can be adopted as follows:

$$\mathrm{MTTF} = \frac{A}{e^{B(\sigma - C)}} \tag{7.1}$$

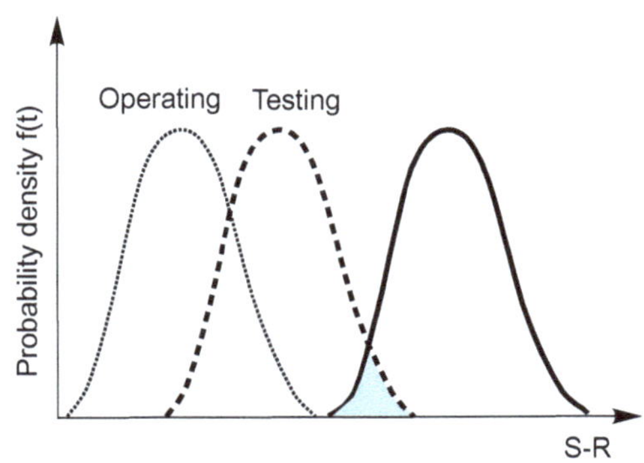

Figure 7.1 Comparison of probability density distribution of failure-related parameters in testing and operating.

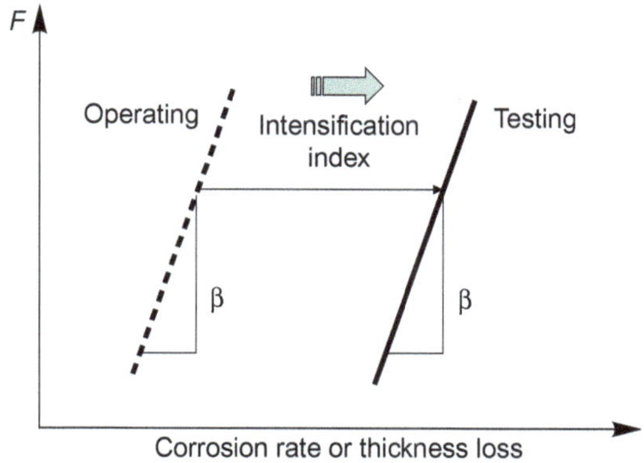

Figure 7.2 Cumulative distribution (Weibull or Gumbel) in accelerated testing and operating.

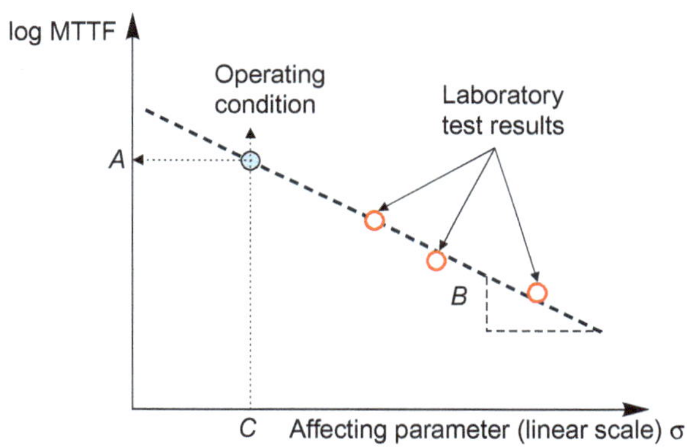

Figure 7.3 Principle of extrapolation from testing results to operating conditions.

where MTTF is generic mean-time-to-failure (or another testing parameter); A, B and C are experimental constants and σ is an affecting parameter. It is worth noting that S.A. Arrhenius (1859–1927) firstly recognized that in accelerated tests the relationship between time and the affecting parameter is logarithmic. Passing to log, Eq. (7.1) becomes:

$$\log(\mathrm{MTTF}) = \log A - B(\sigma - C) \tag{7.2}$$

The plot, log(MTTF) $-\sigma$, in semilogarithmic scale as reported in Fig. 7.3, has the following meaning:

- affecting parameter, σ, varies in the range $\sigma > C$;
- when $\sigma = C$, MTTF $> A$;
- B is slope of the straight line.

MTTF follows a distribution that would be the same for accelerated tests and operating; hence, Weibull module or Gumbel parameter, α, should be equal, as shown in Fig. 7.2.

7.2 Intensification index

Accelerating test results and operating data are linked through the intensification index, as depicted in Fig. 7.2. Corrosion rate increases by accelerating cathodic process or anodic process or both; in practice, accelerated tests are based on the increase of cathodic processes. There are different possibilities, discussed in detail in the following.

7.2.1 Cathodic process is oxygen reduction

To increase corrosion rate when cathodic process is oxygen reduction, three parameters can be changed:

- oxygen content;
- temperature;
- fluid velocity.

which all that parameters influence oxygen limiting current density (see Chapter 2: Uniform Corrosion) through the empirical equation as follows:

$$i_L \cong 10 \cdot 2^{\frac{T-25}{25}} \cdot [O_2] \cdot \left(1 + \sqrt{v}\right) \tag{7.3}$$

where symbols and units are known. Fig. 7.4 shows graphically the effect of the change of affecting parameters.

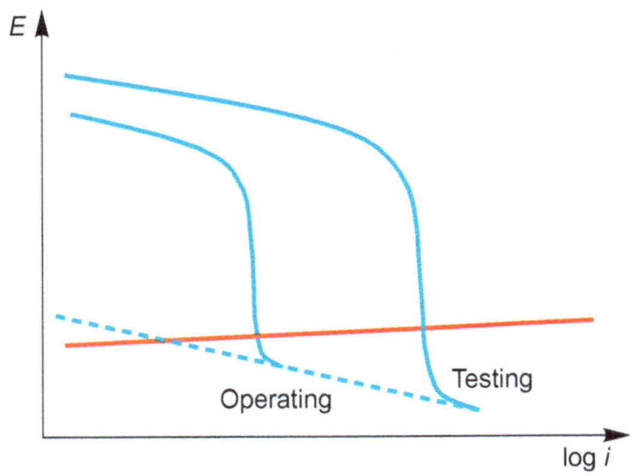

Figure 7.4 Comparison between operating and testing conditions when cathodic process is oxygen reduction.

The intensification index is the ratio between limiting current densities:

$$\text{Index} = \frac{i_{L,\text{test}}}{i_L} = \frac{10 \cdot 2^{\frac{T_{\text{test}}-25}{25}} \cdot [O_2]_{\text{test}} \cdot \left(1+\sqrt{v_{\text{test}}}\right)}{10 \cdot 2^{\frac{T-25}{25}} \cdot [O_2] \cdot \left(1+\sqrt{v}\right)} = 2^{\frac{T_{\text{test}}-T}{25}} \cdot \frac{[O_2]_{\text{test}}}{[O_2]} \cdot \frac{1+\sqrt{v_{\text{test}}}}{1+\sqrt{v}} \tag{7.4}$$

Temperature has an influence for closed, tight circuits, only, because it affects oxygen solubility (refer to Chapter 5: Corrosion in Water, Soil and Air). Testing conditions can be designed for very high intensification index, 100 or higher; for example, by a 10 times increase of oxygen content, a temperature factor up to 2^2 and velocity factor up to 3 (flow rate 4 m/s), intensification index is $10 \cdot 2^2 \cdot 3 \approx 120$. Before designing the accelerating test, it must be verified that anodic process does not change (for instance, no passivation takes place at accelerating conditions) and no scale forms.

7.2.2 Cathodic process is hydrogen evolution

To increase corrosion rate when cathodic process is hydrogen evolution, three parameters can be changed:

- pH;
- temperature;
- fluid velocity.

All these parameters influence hydrogen evolution rate (or current density in mA/m^2), through the general expression:

$$i_{\text{corr,acid sol}} = i_{0,H} \cdot (1+v^2) \cdot 10^{\frac{-0.059\text{pH}-\left(E^0+\frac{0.059}{2}\left(\frac{T+273}{298}\right)\log a_{M^{2+}}\right)}{0.12}} \tag{7.5}$$

where symbols are known. Fig. 7.5 shows graphically the effect of the change of affecting parameters.

The intensification index is given by:

$$\text{Index} = \frac{i_{0,H} \cdot (1+v_{\text{test}}^2) \cdot 10^{\frac{-0.059\text{pH}_{\text{test}}-\left(E^0+\frac{0.059}{2}\left(\frac{T_{\text{tets}}+273}{298}\right)\log a_{M^{2+}}\right)}{0.12}}}{i_{0,H} \cdot (1+v^2) \cdot 10^{\frac{-0.059\text{pH}-\left(E^0+\frac{0.059}{2}\left(\frac{T+273}{298}\right)\log a_{M^{2+}}\right)}{0.12}}} = \frac{(1+v_{\text{test}}^2)}{(1+v^2)} \cdot 10^{\frac{0.059(\text{pH}-\text{pH}_{\text{test}})-\left[\frac{T_{\text{tets}}-T}{298}\right]\frac{0.059}{2}\log a_{M^{2+}}}{0.12}} \tag{7.6}$$

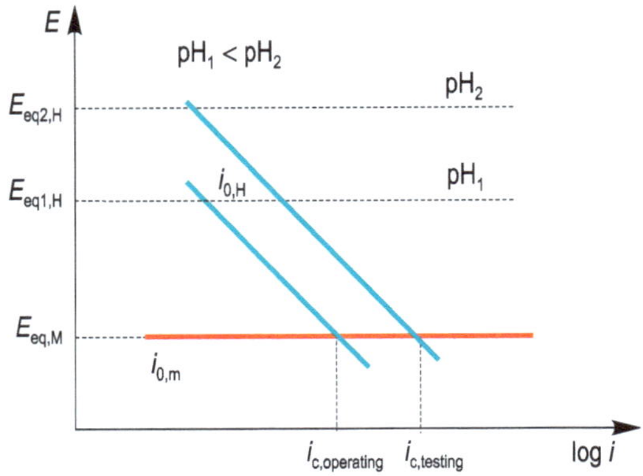

Figure 7.5 Comparison between operating and testing conditions when cathodic process is hydrogen evolution.

Assuming metal ion concentration of 10^{-2} mol/L or higher, a fluid velocity of about 2 m/s as maximum, a temperature increase of 50°C and a pH lowering of 3 units, intensification index is about 200.

7.2.3 Example of intergranular corrosion test

A typical accelerated test is the one for checking susceptibility of intergranular corrosion of stainless steels and nickel-based alloys. Intergranular corrosion occurs when carbides or sigma phase precipitate at grain boundaries. Even in low aggressive environments, a galvanic-type corrosion attack at grain boundary takes place, because carbides and sigma phase are strongly "cathodic" either for hydrogen evolution or oxygen reduction. Although surface area ratio is always the highest, corrosion rate may be relatively low, depending on environment aggressiveness, so there is a need for an accelerated test.

Intensification index depends on process environment, specifically on cathodic process. In Huey test [3], the specimen is immersed in a boiling 65% solution of nitric acid for five periods, each of 48 h. Corrosion rate is calculated from mass loss measurements. If there is no sensitization, i.e. grains are homogeneously passivated, nitric acid maintains passivity and mass loss refers to passivity current with some oxide dissolution. When metal is sensitized, intergranular attack occurs on chromium-depleted zones, which are not passive so that galvanic corrosion occurs between these zones and surrounding passive ones. The cathodic reaction is hydrogen evolution (reduction of nitrate anion does not contribute to the cathodic process); therefore, corrosion rate is given by the cathodic current density of hydrogen evolution reaction, according to the equation:

$$i_{\mathrm{H}} = i_{0,\mathrm{H}} \cdot 10^{\frac{-0.059\mathrm{pH} - \left(E^0 + \frac{0.059}{2}\left(\frac{T+273}{298}\right)\log a_{\mathrm{M}^{2+}}\right)}{0.12}} \cong i_{0,\mathrm{H}} \cdot 10^{\frac{0.4-0.059\mathrm{pH}}{0.12}} \quad (7.7)$$

where $a_{M^{z+}} > 1$ M inside groove and temperature is about 150°C. By taking pH = 0, $i_{0,H} = 10$ mA/m^2, i_H is about 21 A/m^2; hence, intergranular corrosion rate is about 30 mm/y, given by Eq. (3.20) ($\Delta V = 0.1$ V; $(\rho + 0.8) \cong 1$ Ωm).

In process fluid like a neutral aerated water, with 10 ppm of oxygen, intergranular corrosion rate is about 0.7 mm/y, given by Eq. (3.20) ($\Delta V = 0.5$ V; $(\rho + 0.8) \cong 50$ Ωm; $i_L = 0.1$ A/m^2): Accordingly, intensification index is about 40.

7.2.4 Example of test based on thickness loss measurements

When accelerated test is based on thickness loss measurements, as in high temperature and atmospherically exposed testing, the interpretation of test results has to take into account how thickness loss changes with time. Two trends generally match: linear and parabolic; sometimes, it also may fit a logarithmic one. When linear, the prediction of the thickness loss is directly derived from the actualization of the corrosion rate, as above discussed. Intensification index coincides with the one calculated for the corrosion rate.

When, besides the accelerated corrosion rate, thickness loss follows a parabolic law, to calculate the intensification index it is necessary to transform the thickness loss to a mission-time (i.e. time-to-failure or service time) which is, in most cases, the corroding time of the corrosion allowance.

An example helps explain the possible procedure. Let's suppose that we have carried out a laboratory test under accelerated conditions, maintaining same corrosion mechanism (hence, module β of Weibull distribution is the same) obtaining the following data (arbitrary units): 2, 3, 3, 4, 4, 5, 6 after testing time, t_{test}. Based on corrosion mechanism, expected thickness loss is 3 (same arbitrary unit), accordingly mission time (i.e. time-to-failure) is $t_1 = 3/2t_{test}$; $t_2 = t_{test}$; $t_3 = t_{test}$; $t_4 = 3/4t_{test}$; $t_5 = 3/5t_{test}$; $t_6 = 0.5t_{test}$. By plotting on Weibull paper the cumulative distribution, β, δ and γ parameters are obtained. Mission time is then δ if module β exceeds the value 6 or has to be calculated by Eq. (8.19). As thickness loss follows a parabolic law, constant k_{test} can be obtained. The constant k for thickness loss prediction at actual time or in future is obtained by multiplying k_{test} by the square root of time ratio, $(t_{test}/t_{inspection})^{0.5}$. Equivalent procedure applies for a logarithmic relationship.

7.3 Accelerated test with a noble cathodic process

Typical examples are those tests that specify deaerated ferric ion containing solutions. The "new" cathodic process is $Fe^{3+} + e = Fe^{2+}$, which is much more noble than hydrogen evolution. Equilibrium cathodic potential is given by Nernst equation as follows:

$$E_{eq} = E^0 + 0.059\left(\frac{T + 273}{298}\right)\log\frac{a_{Fe^{3+}}}{a_{Fe^{2+}}} \cong E^0 = +0.7 \text{ V SHE} \tag{7.8}$$

because $[Fe^{3+}]/[Fe^{2+}]$ ratio is about unitary. It follows that driving voltage of the corrosion reaction is $[+0.7 - E_{eq,Fe}] \gtrsim 1$ V. This value is much higher than usual driving voltage for corrosion in acidic, aerated solutions.

7.3.1 General corrosion of mild steel

Corrosion reaction is as follows:

$$Fe + 2Fe^{3+} = 3Fe^{2+} \tag{7.9}$$

Corrosion rate is governed by overvoltage and ohmic drop dissipations. Activation overvoltage and ohmic drop are both from very low to negligible; therefore, the relevant dissipation is concentration overvoltage of Fe^{2+} cations, produced by both anodic and cathodic reaction. Therefore, it can be assumed that there is a limiting current density, $i_{L,Fe^{2+}}$, which governs the corrosion rate, which can be approximated by the Fick Law as follows:

$$i_{L,Fe^{2+}} = 2FD\frac{\Delta C}{\Delta x} \tag{7.10}$$

where symbols are known and $2F$ is charge of a mole of ferrous ions. Corrosion rate is one-third of such limiting current density, which can be rounded on the basis of laboratory tests as follows:

$$C_{rate} = 1.2 \cdot \frac{i_{L,Fe^{2+}}}{3} \cong 10 \cdot 2^{\frac{T-25}{25}} \tag{7.11}$$

where corrosion rate, C_{rate}, is in mm/y and T (°C) is the temperature. Intensification index is of the order of magnitude of 10^3, when compared to corrosiveness of aerated solutions, and 10^2, when related to acidic corrosion.

Free corrosion potential coincides with equilibrium potential of active corroding metal that is near −0.4 V SHE for iron, which is in accordance with the proposed mechanism.

7.3.2 General corrosion of stainless steels and Ni-based alloys

7.3.2.1 Low PREN stainless steels

In ferric chloride solution, stainless steels with PREN lower than 30 depassivate thoroughly, then exhibiting same behaviour as mild steel. (Note: Presence of crevices delays the complete depassivation because the crevice macrocell strengthens passivity around it.)

7.3.2.2 High PREN stainless steels and Ni-based alloys

For Ni-based alloys and stainless steels with PREN higher than 40, both showing a pitting potential more noble than redox potential of ferric/ferrous couple, passivity is not destroyed and corrosion behaviour is shown in Fig. 7.6.

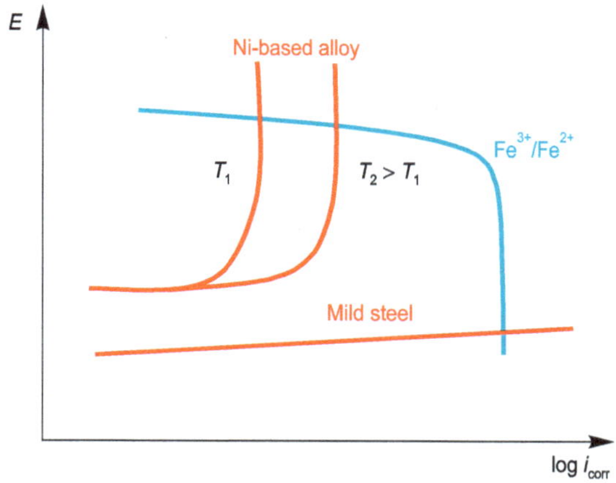

Figure 7.6 Corrosion conditions for mild steel and Ni-based alloys in ferric chloride solution.

Corrosion rate, i_{corr}, coincides with passivity current density, which is function of PREN, chloride concentration, pH and temperature, then given by Eq. (1.25):

$$i_{\mathrm{corr}} = i_{\mathrm{p}} \cong 10^{\left(0.5\cdot\log\left[\frac{\mathrm{Cl}}{36}+1\right]-\frac{\mathrm{PREN}}{13}-\frac{\mathrm{pH}-7}{5}\right)} \cdot 2^{\frac{T-25}{50}} \tag{7.12}$$

where T is the temperature in °C and current density is in $\mathrm{mA/m^2}$. Intensification index coincides with increase temperature parameter, $2^{(T-25)/50}$. Free corrosion potential coincides with redox potential around +0.7 V SHE, which is in accordance with the proposed mechanism.

7.4 Accelerated test for pitting corrosion

Accelerated tests are designed for determining:

- pitting potential;
- PIT (pitting induction time).

For *Critical Chloride Concentration*, there is no need of accelerated test, as it can be determined once fixed the operating conditions.

7.4.1 Pitting potential

According to Eq. (4.4), pitting potential is function of many variables. By taking constant the PIT, which means that once reached the pitting potential, at least a pit starts promptly, and by taking constant all other affecting parameters but one used for the testing, for instance chloride concentration [Cl], pitting potential, E_{pit}, can be extrapolated by the derived equation in accordance with Eq. (7.1):

Table 7.1 Summary of equations used in corrosion testing

	Equations	Parameters
Mean-time-to-failure	$\mathrm{MTTF} = \dfrac{A}{e^{B(\sigma - C)}}$ $\log(\mathrm{MTTF}) = \log A - B(\sigma - C)$	$a_{M^{2+}}$ (mol/L) metal ion concentration A experimental constant B experimental constant C experimental constant [Cl] (ppm) chloride content C_{rate} (mm/y) corrosion rate i_{corr} (A/m^2) corrosion rate [O_2] (ppm) oxygen content PREN = %Cr + 3.3%Mo + 16%N σ affecting parameter T (°C) temperature v (m/s) fluid velocity Y affecting parameter
Intensification index for oxygen reduction	$\mathrm{Index} = 2^{\frac{T_{test}-T}{25}} \cdot \frac{[O_2]_{test}}{[O_2]} \cdot \frac{1+\sqrt{v_{test}}}{1+\sqrt{v}}$	
Intensification index for hydrogen evolution	$\mathrm{Index} = \frac{(1+v_{test}^2)}{(1+v^2)} \cdot 10^{\frac{0.059(pH - pH_{test}) - \left[\frac{T_{tets}-T}{298}\right]\frac{0.059}{2} \log a_{M^{2+}}}{0.12}}$	
Intergranular corrosion test	$i_H \cong i_{0,H} \cdot 10^{\frac{0.4-0.059pH}{0.12}}$	
Corrosion rate of mild steel in ferric chloride solution	$C_{rate} = 1.2 \cdot \frac{i_{L,Fe^{2+}}}{3} \cong 10 \cdot 2^{\frac{T-25}{25}}$	
Corrosion rate of high PREN SS and Ni-alloys in ferric solution	$i_{corr} = i_p \cong 10^{\left(0.5 \cdot \log\left[\frac{Cl}{36}+1\right] - \frac{PREN}{13} - \frac{pH-7}{5}\right)} \cdot 2^{\frac{T-25}{50}}$	
Pitting potential plot	$\log[Y] = \log A - B(C - E_{pit})$	
PIT (pitting induction time) plot	$\log(\mathrm{PIT}) = \log A - B([Cl] - C)$	

$$E_{\text{pit}} \cong f(T, \text{PIN}, v, \text{Cl}, \text{pH}) \quad (7.13)$$

$$\log[Y] = \log A - B(C - E_{\text{pit}}) \quad (7.14)$$

where A, B and C are constants and Y is the affecting parameter (T or v or Cl or pH). By plotting log[Y] vs E_{pit}, pitting potential is extrapolated at operating value of the affecting parameter, Y.

7.4.2 Pitting induction time

Similarly as above, once kept constant the potential, PIT depends on many parameters, which the most influencing one is chloride concentration. By expressing PIT as function of chloride concentration by an Arrhenius-type equation:

$$\text{PIT} = \frac{A}{e^{B([\text{Cl}]-C)}} \quad (7.15)$$

$$\log(\text{PIT}) = \log A - B([\text{Cl}] - C) \quad (7.16)$$

where A, B and C are constants. The plot in semilogarithmic scale, log PIT vs [Cl] allows to extrapolate PIT for operating chloride content. Reference is made to Chapter 4, Pitting and Crevice Corrosion, for more details and case studies.

7.5 Summary

Table 7.1 summarizes the equations used for designing corrosion tests.

References

[1] NACE Standard TM0177. Laboratory testing of metals for resistance to sulphide stress cracking and stress corrosion cracking in H_2S environments. Nace Int. TX; 2005.
[2] NACE Standard TM0284. Evaluation of pipeline and pressure vessel steels for resistance to hydrogen-induced cracking. Nace Int. TX; 2005.
[3] ASTM A262-02. Standard practices for detecting susceptibility to intergranular attack in austenitic stainless steels. ASTM International, PA; 2002.

Statistical Analysis of Corrosion Data

8

Once corrosion data from field inspections or laboratory tests are gathered, there is a need for a proper statistical analysis to obtain, for example minimum/mean/maximum values for the following:

- corrosion rate;
- thickness loss;
- time-to-failure.

It appears evident that in most cases, in particular for localized corrosion, extreme value statistics is the most appropriate. There are also cases where a conclusion must be achieved on the basis on a few data sets. Most often, in design phase, there is a need to evaluate possible data changes; in operating, after a failure or to prevent it, there is the need to determine the most suitable interval of inspections, and so on. In all above cases, extreme value statistics seem necessary.

8.1 Management of corrosion data

As reported in the following chapters, expected corrosion rates can be calculated by adopting models and algorithms derived from corrosion theory. An important question arises: how have these values to be used: as fixed, deterministic value, or mean or even extreme values (minimum or maximum)?

For each corrosion form, the meaning of the calculated value is specified as follows:

- mean;
- maximum or minimum.

These values belong to a distribution; hence, there is a need of confident interpretation. A general procedure might be the following:

- calculated or measured value is the mean;
- for the corrosion form considered, the *Coefficient of Variation*, CV, has to be specified. In general, it is suggested by the experience;
- maximum expected value can be estimated by the application of Gumbel statistics, for a given return period.

8.1.1 Mean or expected value

Mean value, μ, also called *expected value* (from Latin: expectatio), is defined as follows:

Engineering Tools for Corrosion. DOI: http://dx.doi.org/10.1016/B978-0-08-102424-9.00008-2

$$\mu = \int_{-\infty}^{+\infty} X \cdot f(X) dX \tag{8.1}$$

where *f*(*X*) is *probability density function*, *pdf*. For a finite stochastic variable (s.v.), which assumes values v_0, v_1, ..., v_n with probabilities, p_0, p_1, ..., p_n, the following set applies:

$$X = \begin{Bmatrix} v_0 & v_1 & \dots & v_n \\ p_0 & p_1 & \dots & p_n \end{Bmatrix}$$

The mean value, μ, is the sum of all products of each stochastic variable multiplied by the corresponding probability:

$$\mu = v_0 p_0 + v_1 p_1 + \cdots + v_n p_n \tag{8.2}$$

Let be $\mu \equiv E(X)$, some properties are the following:

$$E(aX + b) = aE(X) + b; \quad E(aX + bY) = aE(X) + bE(Y); \tag{8.3}$$

where X and Y are two finite s.v. and a, b are real numbers.

8.1.2 Standard deviation

The *variance*, $\sigma^2(X)$, of a stochastic variable with mean μ is defined for continuous or stochastic variable as follows:

$$\sigma^2 = \int_{-\infty}^{+\infty} (t - \mu)^2 \cdot f(t) dt \tag{8.4}$$

$$\sigma^2(X) = E((X - \mu)^2) \tag{8.5}$$

That is the mean of the square deviation $(X - \mu)^2$. $\sigma(X)$ is called *standard deviation*. The following properties apply:

$$\sigma^2(aX + b) = a^2\sigma^2; \quad \sigma^2(X + Y) = \sigma^2(X) + \sigma^2(Y) \tag{8.6}$$

where X and Y are two finite s.v. and a, b are two real numbers.

8.1.3 Coefficient of variation

The *coefficient of variation* (CV) is a dimensionless number defined as follows:

$$\mathrm{CV} = \frac{\sigma}{\mu} \tag{8.7}$$

where σ is the standard deviation and μ is the mean. It shows the extent of variability with respect to the mean of the population. Most corrosion data show a $\mathrm{CV} < 1$; typically, uniform corrosion has a CV in the range 0.2–0.4 (low variance) and localized corrosion 0.6–0.8 (high variance).

8.2 Extreme value distributions (Gumbel, Weibull)

When extreme values of a phenomenon, although its low probability, are of interest because determining the reliability, for instance, the maximum pit depth or the minimum time-to-failure, specific distributions are used, called *Extreme Value Distributions* and known as *Gumbel distribution* (or double exponential) and *Weibull distribution.* It is worth noting that for a certain target, for example the mean-time-to-failure (MTTF) for stress corrosion cracking (SCC), the pertaining law is often unknown; instead only extreme values are of interest, possibly derived from a small set of data.

8.2.1 The Gumbel distribution

E.J. Gumbel (1891–1966) introduced a double exponential *pdf* as function of two parameters, α and λ, respectively, for *scale parameter* and *location parameter*, which are as follows:

$$f_{\mathrm{G}}(x) = \frac{1}{\alpha} \mathrm{e}^{\left[-\left(\frac{x-\lambda}{\alpha}\right) - \mathrm{e}^{-\left(\frac{x-\lambda}{\alpha}\right)}\right]} \tag{8.8}$$

This *pdf* has an asymmetric shape, shifted towards positive values of the normalized variable $[(x - \lambda)/\alpha]$. Cumulative function, $F_{\mathrm{G}}(x)$, or hazard function, is as follows:

$$F_{\mathrm{G}}(x) = e^{-e^{-\frac{x-\lambda}{\alpha}}} \tag{8.9a}$$

easily obtained as median rank, for example $F(i) = i/(N+1)$ (see appendix). By plotting in semi-logarithm scale the function (8.9a):

$$y = -\ln[-\ln F_G(x)] = \frac{x-\lambda}{\alpha} \tag{8.9b}$$

a straight-line is obtained; λ and α are obtained graphically ($\lambda = x$ for $y = 0$ and α is slope of the line). Mean value and standard deviation are function of parameters α and λ:

$$\mu = \lambda + 0.58\ \alpha \tag{8.10}$$

$$\sigma = 1.28\ \alpha; \tag{8.11}$$

where 0.58 is the Euler's constant. Maximum value can also be calculated once fixed the confidence or *Return Period*, $1/T$, by the following most used equation:

$$x_{max} = \lambda + 4.6\ \alpha \tag{8.12}$$

where constant 4.6 applies to a return period of $1/T = 101$, i.e. for a cumulative function value of 0.99.

8.2.1.1 Gumbel analysis of general corrosion data

When treating uniform corrosion data, namely corrosion rate or thickness loss, normal distribution can be the first choice. Experimental data belong to the interval $\pm 2\ \sigma$ from mean value, μ, and the maximum expected value can be taken as $\mu + 2\ \sigma$, where the standard deviation, σ, is obtained from the coefficient of variation, CV. For example, by assuming for uniform corrosion a CV equals to 0.2, the maximum value is $x_{max} \cong 1.4\ \mu$.

When generalized corrosion is roughly non-uniform, Gumbel statistics seems to be more appropriate. In this case, by taking CV as 0.4, maximum expected value can be calculated by the Gumbel statistics as follows:

$$\begin{aligned} \sigma &= CV \cdot \mu = 0.4\ \mu; \\ x_{max} &= \mu \cdot (1 + 3.14\ CV) = 2.25\ \mu \end{aligned} \tag{8.13}$$

8.2.1.2 Gumbel analysis of localized corrosion data

For the analysis of localized corrosion data, Gumbel's basic idea that he formulated in 1958 appears quite appropriate. As Gumbel argued, if only extreme values are of interest the study can be restricted to highest indications, which are known, regardless the 'normal' trend, which is generally not known; furthermore, if independence is verified for values distributed in fixed intervals, then the trend of distribution of extreme values is found regardless the distribution of the variable. In short, let $f(x)$ the probability that value x happens in a specified interval, if Y is the maximum value that x matches in that interval, the variable Y is also a distribution like $f(x)$, provided, as it is generally accepted, that:

- mechanism and conditions do not change (past, present and future are equivalent). This involves that scale factor, α, which represents the dispersion of data ($1/\alpha$ is the slope of the straight line) is constant for each set of specimens, even taken at different time;
- data are gathered from independent events, therefore the process is considered memoryless and each sampling is considered independent when taken in different periods.

However, location factor, λ, which is time dependent, increases as phenomenon proceeds and kinetics can establish a temporal law.

From gathered data, α and λ are obtained and maximum value can be calculated. An exercise is reported in appendix.

8.2.2 The Weibull distribution

W. Weibull (1887–1979) introduced a *pdf* defined by three parameters, which are as follows:

$$f(t) = \frac{\beta}{\delta}\left(\frac{t-\gamma}{\delta}\right)^{\beta-1} e^{-\left(\frac{t-\gamma}{\delta}\right)^{\beta}} \tag{8.14}$$

where β is the *shape parameter*, also called *Weibull module*; δ is the *scale parameter*; γ is the *threshold parameter* (often taken as zero when starting point is the origin). Reliability, failure and hazard functions are given respectively:

$$R(t) = \int f(t)dt = e^{-\left(\frac{t-\gamma}{\delta}\right)^{\beta}} \tag{8.15}$$

$$F(t) = 1 - e^{-\left(\frac{t-\gamma}{\delta}\right)^{\beta}} \tag{8.16}$$

$$\lambda(t) = \frac{f(t)}{R(t)} = \frac{\beta}{\delta}\left(\frac{t-\gamma}{\delta}\right)^{\beta-1} \tag{8.17}$$

Weibull distribution is very versatile and works with extremely small number of samples. The Weibull module, β, is particularly significant and may provide an interpretation of the process (or the physics of the failure). With less than 20 samples, the Weibull is the best choice, and therefore, best practice. Furthermore, Weibull distribution includes or approximates other distributions for the following shape parameter, β, values:

Exponential	1
Rayleigh	2
Lognormal (approximate)	2.5
Normal	3.44

The slope, β, also indicates which class of event is present:

- $\beta < 1$ indicates that the rate of the event decreases over time. This happens if there is significant 'infant mortality', as in case of pitting or defective items early failing;
- $\beta = 1$ indicates a constant rate. This might suggest that random external causes take place;
- $\beta > 1$ indicates that the event rate increases with time. This happens if there is an 'ageing' process or parts that are more likely to fail as time goes on;
- $\beta > 6.0$, μ and δ approximately coincide.

Scale parameter, δ, changes the scale of the *pdf* along the random variable axis (most usually the time, for example changing the scale of time from days to months or from hours to days); instead, it does not change the shape of the distribution. A property of δ is that it represents the time (or other variable value) by which 63.2% of the population fails, regardless the module β.

Threshold parameter or *location parameter*, γ, is the minimum value of the random variable. In most applications this parameter is taken as zero, therefore we obtain the *Weibull-two-parameter-distribution*.

To plot Weibull distribution, normalized variable, z, is often used:

$$z = \beta \cdot [\ln(x-\gamma) - \ln \delta]; \quad F(z) = 1 - e^{-e^{z}} \tag{8.18}$$

$F(z) = 0.632$ corresponds to $z = 0$; this point is often used as starting point to determine β graphically. If, by putting $\gamma = 0$, the plot is not a straight line, then $\gamma > 0$ is tentatively used to obtain the straight line. If this tentative fails, another distribution should be used (for instance log-normal). γ, δ and β are generally obtained graphically; β can be calculated analytically by means of the following equation:

$$\beta = \frac{\ln\left\{\ln\left[\frac{1}{1-F}\right]\right\}}{\ln(X-\gamma) - \ln \delta} \tag{8.19}$$

where F is the cumulative as median rank and X the random variable. μ or MTTF, and standard deviation, σ, are given by the following equations:

$$\mu = \text{MTTF} = \gamma + \delta\, \Gamma\left(\frac{1}{\beta} + 1\right) \tag{8.20}$$

$$\sigma = \delta \sqrt{\Gamma\left(\frac{2}{\beta} + 1\right) - \Gamma^2\left(\frac{1}{\beta} + 1\right)} \tag{8.21}$$

where Γ is the *Gamma function*, which can be calculated as reported in Table 8.1. For $\beta > 6.0$, Gamma function approximates 1.

8.3 Time-dependent data

8.3.1 *Hazard and reliability functions*

This is the typical case of analysis of inspection data, which vary with time. As an example, annual inspection of heat exchanger tubes reports increasing number of failures. At each annual inspection, if n_i is number of failures on a total tube

Table 8.1 Values of gamma function

k	$\Gamma(\frac{1}{k}+1)$	k	$\Gamma(\frac{1}{k}+1)$	k	$\Gamma(\frac{1}{k}+1)$	k	$\Gamma(\frac{1}{k}+1)$
0	∞	1	1.00	2.0	0.886	3.0	0.894
0.1	10!	1.1	0.965	2.1	0.886	3.1	0.894
0.2	5!	1.2	0.941	2.2	0.886	3.2	0.896
0.3	9.261	1.3	0.923	2.3	0.886	3.3	0.897
0.4	3.323	1.4	0.911	2.4	0.886	3.4	0.898
0.5	2.00	1.5	0.903	2.5	0.887	3.5	0.900
0.6	1.505	1.6	0.897	2.6	0.888	3.6	0.901
0.7	1.266	1.7	0.892	2.7	0.889	3.7	0.902
0.8	1.133	1.8	0.889	2.8	0.890	3.8	0.904
0.9	1.052	1.9	0.887	2.9	0.892	3.9	0.905

number, N, the ratio n_i/N is failure rate at year ith. After m inspections (time $t = m$ years), the cumulative hazard function, H, is calculated as sum of failure rates:

$$H(t) = \sum_{i=0}^{m} \frac{n_i(t)}{N} \tag{8.22}$$

Reliability, R, at year m, after m annual inspections, is:

$$R(m) = \mathrm{e}^{-H(m)} = \mathrm{e}^{-\sum_{i=0}^{m} \frac{n_i(t)}{N}} \tag{8.23}$$

Then, cumulative failure function is as follows:

$$F(t) = 1 - R(t) = 1 - \mathrm{e}^{-\sum_{i=0}^{m} \frac{n_i(t)}{N}} \tag{8.24}$$

From the plot $\ln F - t$, lifetime is extrapolated at maximum allowed tube failures. In appendix an example is reported.

8.4 Sampling

Often for testing or inspection survey, it is not possible to examine all elements or components for economic or even for technical convenience; in other words, a sampling is necessary.

As the results obtained by testing a limited number of components, indeed represent the properties of the sample and can be extended to the whole population by inference with a certain level of confidence, only. It is also said that sampling must be representative and free from bias.

To achieve this goal, sampling must follow a correct procedure to minimize the so-called 'sampling error'. The latter can derive from an inherently casual variation that is minimized by selecting, randomly, a minimum number of samples; or due to a wrong procedure because affected by a bias. To avoid the bias, the sampling procedure has to be based on a principle of pure casual selection so that each member (component or element) has the same probability to be part of the sample.

By excluding the non-probabilistic method carried out on a conscious selection, which is obviously not correct, the *sampling* can be as follows:

- *Random*, by a casual extraction of elements on previous numbered population. This sampling procedure, although the most correct, is often not practical;
- *Systematic*, by selecting elements by regular intervals, for instance 1 out of 10 after numbering. It can be influenced by external constraints, nevertheless it assures good randomness;
- *Stratified*, when population can be divided into homogeneous groups, each one characterized by a common property which is the objective of the survey. In each group, the sampling follows a random or systematic procedure;
- *Cluster*, when for ease population is divided into clusters. In each cluster, the sampling follows a random or systematic procedure. This sampling can be affected by a higher error than that of other methods.

8.4.1 Some parameters

In the following, some important parameters affecting sampling are briefly highlighted.

Standard error, SE, of a mean is given by:

$$\mathrm{SE} = \frac{\sigma}{\sqrt{N}} \tag{8.25}$$

where σ is *standard deviation* and N population numerosity.

In sampling, the following parameters are used:

- *prevalence*, p, or frequency of occurrences (for instance number of occurrences of thickness loss exceeding a given value);
- *variance* (σ^2) given by (σ^2) = $p(1-p)/n$, where n is the sample size;
- *Correction factor*, CF = $(N-n)/(N-1)$, where N and n are population and sample size, respectively;
- SE, given by SE = $[(p)\times \mathrm{CF}]^{1/2} \cong (p)^{1/2}$;
- *Estimate variance*, given by EV = $p \times$ SE.

Assuming for example, prevalence $p = 0.35$ which means that 35% of measurements fits with established criterion (for instance thickness loss exceeding a given value); a population of 500 elements and a sample of 40 elements, results are the following: *variance* (σ^2) = 0.00569; CF $\cong$ 1; SE = 0.0754 = 7.54%; EV = 35% $\pm$ 7.54.

Confidence intervals represent the probability to find the value in a range of the population and are defined as follows:

- *mean* ± 1 *SE*, or 68% *confidence interval*;
- *mean* ± 2 *SE*, or 95% *confidence interval*;
- *mean* ± 3 *SE*, or 99% *confidence interval*.

For instance, if *confidence interval* is 95% for the above example, lower limit is $0.35 - (2 \times 0.0754) = 0.199 = 19.9\%$ and upper limit is $0.35 + (2 \times 0.0754) = 0.508 = 50.1\%$. In summary, with a confidence of 95%, the percentage of occurrences in the population is in the interval 20% to 50%.

8.4.2 Determination of sample size

It should be noted that for vast population, the optimal size for sampling is independent from numerosity of the population. Without discussing this point, most used relationships for determining the minimum (or optimal) size, n, of the sample for a vast population, N, are the following:

$$n = \frac{k^2 P_{\text{exp}}(1 - P_{\text{exp}})}{E^2} \tag{8.26}$$

where P_{exp} is expected confidence (desired), E absolute error and k is a constant which depends on confidence (1 for confidence 68%, 1.96 confidence 95% and 2.576 for confidence 99%);

$$n = \frac{\ln a}{\ln p} \tag{8.27}$$

used for high numerosity population, where a is confidence and p prevalence ($>5\%$);

$$\alpha = nP^{n-1} - (n-1)P^n \tag{8.28}$$

or Wilks equation (by S.S. Wilks, 1906–64) that is valid regardless the type of distribution. P is expected probability that occurrence falls in a fixed interval with confidence $1 - \alpha$. For instance, with confidence 95% to determine the limits for 90% of the population, $n = 45$.

8.5 Summary

Table 8.2 summarizes all equations used for statistical analysis.

Table 8.2 Summary of equations used for statistical analysis of corrosion data

	Equations	Parameters
Mean value of a *pdf*	$\mu = \int_{-\infty}^{+\infty} X \cdot f(X) dX$	$f(X)$ is *pdf* $v_0, v_1, \ldots, v_n$ values of s.v. $p_0, p_1, \ldots, p_n$ associated probabilities σ Standard deviation μ Mean σ Standard deviation
Mean value of stochastic variable	$\mu = v_0 p_0 + v_1 p_1 + \cdots + v_n p_n$	
Variance and standard deviation	$\sigma^2 = \int_{-\infty}^{+\infty} (t-\mu)^2 \cdot f(t) dt$	
	$\sigma^2(X) = E((X-\mu)^2)$	
Coefficient of variation, CV	$CV = \frac{\sigma}{\mu}$	
Gumbel distribution	$F_G(x) = e^{-e^{-\frac{x-\lambda}{\alpha}}}$	*Return Period,* $1/T = 0.99$ $\mu = \lambda + 0.58\alpha$ $\sigma = 1.28\alpha$ CV = 0.2 for uniform, Normal Distr. CV = 0.4 for general, Gumbel Distr.
	$y = -\ln[-\ln F_G(x)] = \frac{x-\lambda}{\alpha}$	
Maximum value in Gumbel distribution ($1/T = 0.99$) Uniform corrosion General corrosion	$x_{max} = \lambda + 4.6\alpha$ $x_{max} \cong 1.4\alpha$ $x_{max} \cong 2.25\mu$	
Weibull distribution	$F(t) = 1 - e^{-\left(\frac{t-\gamma}{\delta}\right)^{\beta}}$	β Shape parameter or Weibull module δ Scale parameter γ Threshold parameter (often zero) $z = \beta\,[\ln(x-\gamma) - \ln \delta]$; γ Threshold parameter (often zero) F Cumulative distribution Γ Gamma function
	$F(z) = 1 - e^{-e^{z}}$	
Weibull module, β	$\beta = \frac{\ln\left\{\ln\left[\frac{1}{1-F}\right]\right\}}{\ln(X-\gamma) - \ln \delta}$	
Mean or MTTF in Weibull distribution	$\mu = MTTF = \gamma + \delta\, \Gamma\left(\frac{1}{\beta} + 1\right)$	
Standard deviation in Weibull distribution	$\sigma = \delta \sqrt{\Gamma\left(\frac{2}{\beta} + 1\right) - \Gamma^2\left(\frac{1}{\beta} + 1\right)}$	

Median rank proposed by Fisher	$F(i)=\frac{1}{1+\frac{N+1-i}{i}F_{conf,m,n}}$	N Number of samples $F_{conf,m,n}$ Characteristic distribution available in tables
Median rank most used equation (for a confidence of 50%)	$F(i)=\frac{i-0.3}{N+0.4}$	N Number of samples
Median rank simplified equation	$F(i)=\frac{i}{N+1}$	N Number of samples
Normal distribution – failure function (normalized)	$F(x)=\phi(z)=\frac{1}{\sqrt{2\pi}}\int_{-\infty}^{z} e^{\frac{-z^2}{2}}dz$	z Normalized variable $\phi(z)$ Hastings equation
	$\phi(z)=1-0.5\left(1+0.0499z+0.0211z^2+0.0033z^3+0.000038z^4+0.000049z^5+0.0000054z^6\right)^{-16}$	
Exponential distribution	$f(t)=\lambda^{*}e^{-(\lambda^{*}t)}$	λ^{*} Failure rate (or hazard function) constant t Variable (in general is time)
Log-normal distribution	$f(x)=\frac{1}{\sigma x\sqrt{2\pi}}e^{-\left[\frac{1}{2}\left(\frac{\ln x-\ln\mu}{\sigma}\right)^2\right]}$	$x=\ln z$ σ Standard deviation $\ln\mu$ Mean value
Sampling size (N vast population)	$n=\frac{k^2P_{exp}(1-P_{exp})}{E^2}$	a confidence E absolute error k constant depending on confidence n number of member of a sample P_{exp} is expected confidence (desired) P expected probability p prevalence ($>5\%$)
	$n=\frac{\ln a}{\ln p}$	
	$\alpha=nP^{n-1}-(n-1)P^{n}$	

8.6 Appendix

8.6.1 Median rank and return period

To approximate the cumulative distribution, R. Fisher (1890–1962) proposed the equation of *median rank*:

$$F(i) = \frac{1}{1 + \frac{N+1-i}{i} F_{\mathrm{conf},m,n}} \tag{8.29}$$

where i is the ordinate number of test, $F_{\mathrm{conf},m,n}$ is the characteristic distribution proposed by Fisher available in tables, $m = 2(N + 1 - i)$, $n = 2i$. The median rank for a confidence of 50% is often approximated to:

$$F(i) = \frac{i - 0.3}{N + 0.4} \tag{8.30}$$

or more generally to the simplified one:

$$F(i) = \frac{i}{N + 1} \tag{8.31}$$

From median rank, the following is obtained:

$$\frac{N+1}{N+1-i} = \frac{1}{1-f} = T \tag{8.32}$$

where T is the *return period* which has the meaning of number of samples, N, to comply a confidence f. For example, to achieve a confidence, f, of 90%, at least 11 samples are required.

8.6.2 The normal distribution

This is the most famous distribution that is characterized by a symmetrical *pdf* around a mean value, μ, which is also the most probable one. In corrosion, the variable could be corrosion rate, thickness loss, time-to-failure, potential and others. Its most often form is function of mean value, μ, and standard deviation or variance, σ:

$$f(x) = \frac{1}{\sigma\sqrt{2\pi}} \mathrm{e}^{-\frac{(x-\mu)^2}{2\sigma^2}} \tag{8.33}$$

Curve has a shape of a bell with two flexes at $\pm\sigma$ from mean value, μ, where σ is variance. The cumulative function, $F(x)$, is the following:

$$F(X) = \frac{1}{\sigma\sqrt{2\pi}} \int_{-\infty}^{X} \mathrm{e}^{-\frac{(x-\mu)^2}{2\sigma^2}} dx \tag{8.34}$$

and represents, when expressed as function of time, the failure function, $F(t)$. *Reliability function*, $R(t)$, is $1 - F(t)$, and *hazard function*, $\lambda(t)$, is the *pdf*-to-$R(t)$ ratio. By substituting the ratio, $[(t - \mu)/\sigma]$, with the normalized variable, z, the normalized form is obtained, used to ease calculation with a spreadsheet:

$$F(x) = \phi(z) = \frac{1}{\sqrt{2\pi}} \int_{-\infty}^{z} e^{\frac{-z^2}{2}} dz \tag{8.35}$$

Variable, x, is reported on abscissa in a linear scale $[z = (x - \mu)/\sigma]$ and on ordinate the cumulative obtained by the *median rank*. Hastings equation is often used:

$$F(z) = 1 - 0.5\Big(1 + 0.0499z + 0.0211z^2 + 0.0033z^3 + 0.000038z^4 + 0.000049z^5 + 0.0000054z^6\Big)^{-16} \tag{8.36}$$

The straight line is obtained easily taking into account that:

- when $z = 0$, then $F(z) = 0.5$ and $x = \mu$;
- when $z = 1$, $F(z) = 0.84$ and $x = \mu + \sigma$.

The normal distribution is used for time-dependent phenomenon as aging, fatigue and uniform corrosion. Its use is easy and suitable for prediction rather than for data analysis.

8.6.2.1 Example of the use of normal distribution

We want to calculate the reliability of boiler tubes (subject to generalized corrosion at waterside) after 5-year exposure. Experience showed that mean time-to-failure, μ, is 25 y (i.e. thinning exceeds corrosion allowance) and variance, σ, is 5 y. Corresponding z value is -4 $[z = (t - \mu)/\sigma]$ and from Hastings equation $F(z) < 0$ and reliability is almost 1. At year 20, $F(z)$ is 0.16 then reliability is 0.84; the probability of a failure between year 20 and year 30 is 0.68 (given by $F(30) - F(20)$). Failure rate, $\lambda(t)$, given by the ratio *pdf/reliability*, is 2.7×10^{-5} at year 5 and 0.3 at year 30. The meaning is as follows: if boiler contains 10,000 tube equivalents (number of unitary length, for instance 0.33 m or three per metre) at year 5 the failure probability is less than one unit in total, while at year 30, only about 1600 tubes survive ($F(30) \cong 0.84$), with 30% probability that at least one fails.

8.6.3 The exponential distribution

Most adopted in reliability analysis and also used for behaviour of materials. It fits when failure rate is constant. Then, *pdf* is an exponential equation as follows:

$$f(t) = \lambda^* \mathrm{e}^{-(\lambda^* t)} \tag{8.37}$$

where λ^* is failure rate or the *hazard function* that is constant:

$$\lambda(t) = \frac{f(t)}{R(t)} = \frac{\lambda^* \mathrm{e}^{-(\lambda^* t)}}{\mathrm{e}^{-(\lambda^* t)}} = \lambda^* \tag{8.38}$$

As failure rate, λ, is constant, the probability of a failure (i.e. a corrosion occurrence) is independent from time or also said process without memory (neither infant failure nor aging effect). MTTF is $1/\lambda^*$, also constant. Often, for material behaviour, a threshold parameter, γ, is considered ($\mathrm{MTTF} = \gamma + 1/\lambda^*$) that means that failure occurs after some time, only. An example of exponential distribution seems to be the initiation time of localized corrosion, i.e. when a passive film breakdowns locally as reported in Ref. [1].

8.6.3.1 Example of the use of exponential distribution

Let's assume that failure of galvanic anodes in seawater, caused by passivation or intergranular attack, follows an exponential distribution. For bracelet-type anodes, used to protect a submarine pipeline, failure rate gained from previous experiences, is 10 in 10^6 h ($10/10^6\ \mathrm{h} = 10^{-5}\ \mathrm{h}^{-1}$ or $10/114.2\ \mathrm{y} = 0.088\ \mathrm{y}^{-1}$). What is the reliability at year 10? The calculation is as follows:

$$R(10) = \exp(-10 \times 0.088) = 0.41; \quad R(20) = \exp(-20 \times 0.088) = 0.17$$

$\mathrm{MTTF} = 1/\lambda^* = 1/0.088 = 11$ y. This means that if total anode is 10,000 within the service life of 20 y, about 8300 anodes can suffer of passivation or excessive consumption rate, and mean time to failure is 11 y. It could be also calculated that *mission time* for a minimum reliability of 0.9 is 1.2 y [mission time = ln $(R)/\lambda$]. This result can be interpreted as follows: To assure the specified reliability, almost a yearly check has to be done, which means that every year there is a zeroing.

8.6.4 The log-normal distribution

When variable is a logarithm ($x = \ln z$), *pdf* is given by:

$$f(x) = \frac{1}{\sigma x \sqrt{2\pi}} \mathrm{e}^{-\left[\frac{1}{2}\left(\frac{\ln x - \ln \mu}{\sigma}\right)^2\right]} \tag{8.39}$$

where σ and μ are standard deviation and mean value, respectively. This distribution represents phenomena with pdf shifted on initial time. Typically, it represents the time-to-failure for SCC.

8.6.5 Example of extreme value distribution

8.6.5.1 Localized corrosion

A lab testing was designed to check corrosion propagation occurrence after pitting initiation on a stainless steel component. Aim of testing was as follows:

- reproduce pitting initiation in a solution with high chloride content (10,000 ppm) on a sample of specimens in a sufficient number for a statistical analysis, 10 for each sampling for a total of 40 specimens;
- after pitting initiation, 10 specimens were randomly sampled as control set;
- every 15 day exposure, a set of 10 specimens was removed and analysed (depth of attack was measured);
- statistical analysis carried out by means of Gumbel distribution.

The main result of the application of Gumbel distribution is the estimation of the extreme value, which has the meaning of the maximum value existing among the whole population with a confidence usually taken as 99%. This extreme value can be used for deciding the best date for the future inspection. It was proceeded as follows: each 10 specimen data set is ranked by increasing depth of attack, to which median rank function $[F(i) = i/(N + 1)]$ is calculated to obtain the Gumbel plot, as reported below in Fig. 8.A1. From the graph, α and λ are determined, which allow to estimate the maximum expected value.

8.6.5.2 Tank bottom inspection

This case study refers to an inspection on a tank bottom which consisted on the measurement, by sampling, of the residual thickness. By applying a random sampling (see section 8.4) 10 sectors of the tank bottom were selected. For each sector, maximum thickness loss was measured, then ordered and associated to the cumulative distribution estimated by the median rank, as reported below:

Sector	**1**	**2**	**3**	**4**	**5**	**6**	**7**	**8**	**9**	**10**
Thickness loss (mm)	5.9	6.0	6.1	6.2	6.3	6.6	6.8	7.0	7.3	7.4
Cumulative distribution $F(i)$	0.09	0.18	0.27	0.36	0.45	0.55	0.64	0.73	0.82	0.91

By applying the Gumbel distribution, maximum value is easily extrapolated, having the meaning of the maximum existing value on the whole population with a certain confidence, for example 99%. Parameters α and λ are obtained

Rank	Max pit depth (μm)	$F(x)$	$y = [-\ln(-\ln F)]$
1	120	0.09	−0.87
2	163	0.18	−0.53
3	240	0.27	−0.26
4	240	0.36	−0.01
5	250	0.45	0.24
6	270	0.55	0.50
7	309	0.64	0.79
8	320	0.73	1.14
9	362	0.82	1.61
10	477	0.91	2.35

$\alpha = 110$; $\lambda = 225$; and $x_{max} = 700$ μm

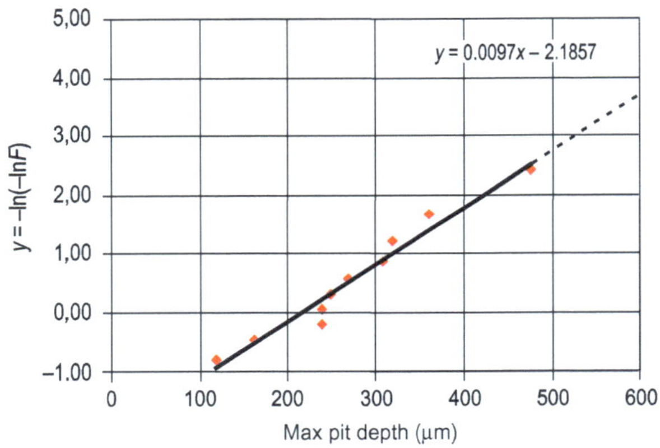

Figure 8.A1 Example of the application of Gumbel extreme value statistics.

($\alpha = 2.34$ mm; $\lambda = 4.84$ mm); maximum thickness loss with confidence 0.99 is 15.6 mm. Mean value is 6.2 mm and standard deviation is 3 mm.

By applying the Weibull distribution, the mean thickness loss is calculated from parameters γ, δ and β given by the plot on Weibull paper or by the relationship (8.20 and 8.21). Shape parameter, β, is 12.8; mean, δ, is $\cong$6.7 mm.

8.6.5.3 Tank bottom inspection

Similarly to the previous case study, the inspection of a tank bottom consisted on residual thickness measurements carried out on 15 sectors selected by random sampling of the tank bottom of 314 m^2 surface area after 10 y from commissioning. Results are shown as follows:

Sector	1	2	3	4	5	6	7	8	9	10	11	12	13	14	15
Median rank (%)	4.5	11	17.5	24	30.5	37	43.5	50.0	56.5	63	69.5	76.0	82.5	89	95.5
Max thickness loss (mm)	1.25	1.6	1.9	2.0	2.1	2.2	2.5	2.5	2.8	2.9	3.1	3.1	3.3	3.4	4.1

By applying Weibull distribution, it results: $\gamma = 0$; $\delta = 2.9$; $\beta = 3.5$; mean value 2.6 (=2.9 × 0.9). The empirical temporal law of thickness loss, Δs, with time is as follows:

$$\Delta s = k\sqrt{t} \tag{8.40}$$

hence, it is possible to calculate the constant k which is 0.82 mm/y$^{0.5}$ (2.6/10$^{0.5}$). As an exercise, it can be estimated the expected mean thickness loss after further 5 years, which is 3.2 mm, then giving a further thickness loss of 0.6 mm (i.e. given by 3.2 − 2.6). Based on this expectation, it was decided to carry out the next inspection after 10 y.

8.6.6 Lifetime calculation

From inspection data, the lifetime can be forecasted. Assuming an exponential distribution of gathered data set, after n inspections a thickness-loss-to-time plot can be used for extrapolating the expected lifetime. For example, the following data are obtained after three inspections:

- 20% of surface, corresponding to 5 sectors, is inspected at three successive time, t_1, t_2 and t_3, obtaining n data for a total of $N = 5n$ measurements. Measurements exceeding a prefixed thickness loss, Δs_j, defined as follows: $\Delta s_j{}^* = t_j$ CA/DL where CA is corrosion allowance, DL, design life, j is index of inspection time, t_j, with $t_j < \text{DL}$;
- failure rate, λ_j, at time j, is ratio between number of measurements n_j for $\Delta s_j > \Delta s_j{}^*$ and total number, N: $\lambda_j = \frac{n_j}{N}$;
- cumulative Hazard function, H, is $H(t) = \sum_{t_j \le t} \frac{n_j}{N}$;
- reliability $R(t)$ is given by $R(t) = e^{-H(t)}$ and cumulative failure function is $F(t) = 1 - R(t)$;
- from the plot ln $F - t$, lifetime is extrapolated at maximum thickness loss equals to corrosion allowance, CA.

For this case study, Weibull distribution could also be used, although less practical for the extrapolation because of the logarithmic scale.

Reference

[1] Kowaka M, Tsuge H, Akashi M, Matsumura K, Ishimoto H. An introduction to life prediction of industrial plant materials—application of extreme value statistical method for corrosion analysis. New York: Allerton Press; 1994.

Further reading

Gumbel EJ. Statistics of extremes. New York: Columbia University Press; 1958.

Kottegoda NT, Rosso R. Statistics, probability and reliability for civil and environmental engineers. New York: McGraw-Hill International Editions; 1998.

Glossary

A

Accelerate Corrosion Testing Corrosion test carried out by imposing test parameters to reduce substantially its duration

Acidic corrosion Corrosion attack in acidic solutions for which cathodic process is hydrogen evolution

Active (1) Expresses a high tendency of a metal to react. (2) A state of a metal that is corroding

Active metal A metal which is active and has a tendency to react (or corrode)

Active–passive metal A metal which can behave in either active or passive state

Aerobic Presence of air or oxygen as dissolved gas

Amphoteric metal A metal that is susceptible to corrosion in both acid and alkaline environments

Anaerobic Free of air or oxygen

Anaerobic environment An environment free of oxygen or air which supports the growth of anaerobic bacteria

Anion A negatively charged ion

Anode The electrode on which oxidation takes place releasing electrons

Anodic current Current flowing across the metal–electrolyte interface from the metal surface to the electrolyte, causing an anodic polarization

Anodic polarization Or anodic overvoltage. The change of the electrode potential in the noble (positive) direction caused by an anodic current

Anodic protection Electrochemical technique to protect active–passive metals by supplying an anodic current

Austenitic stainless steel A stainless steel with austenitic microstructure at room temperature

B

Backing steel Carbon or low-alloy steel that is cladded with a corrosion resistant alloy

Barrier effect The property of coatings or layers with low permeation of liquids and gases; it increases corrosion resistance

BEM Boundary Element Method. A numerical technique for solving field equations based on imposing the constraints on only anodic and cathodic surface

Bimetallic corrosion (see Galvanic Corrosion)

Bimetallic coupling (see Galvanic Coupling)

Biofilm A formation of a bacteria-enriched layer on a metal surface immersed in seawater

Buffer A substance that prevents pH changes when present in the electrolyte

C

Calcareous deposit A calcium carbonate scale formed in hard freshwater or deposited in seawater when CP is applied

Carbon steel Steel containing basic elements only, as carbon and manganese

Cation A positively charged ion

Cathode The electrode on which a reduction reaction takes place by acquiring electrons

Cathodic current Current flowing across the metal–electrolyte interface from the electrolyte to the metal surface, causing a cathodic polarization

Cathodic polarization Or cathodic overvoltage. The change of the electrode potential downwards caused by a cathodic current

Cathodic prevention (CPrev) Electrochemical technique as Pietro Pedeferri (1938–2008) named to protect passive metals from pitting consisting in applying a cathodic current before passive film breakdown

Cathodic protection (CP) Electrochemical technique to protect metals, consisting in lowering the metal potential by supplying a cathodic current

CP by immunity CP condition which brings metals to immunity

CP by passivity CP condition which brings and maintains metals to passive state

Cavitation Damage to a metallic material under conditions of severe turbulent flow

Cell (see electrochemical cell and macrocell)

Chemical equivalent Atomic or molecular mass of an ion divided by its valence

Coating efficiency (see porosity)

Clad, cladding material Layer of a corrosion resisting metal used as corrosion prevention for vessels and pipelines; backing metal is typically carbon and low-alloy steel

Coefficient of variation It is defined as the ratio between standard deviation and mean. It is normally low for uniform corrosion and high for localized one

Computer modelling It is a technique using computer software for solving field equations based on FEM and BEM. Widely used in CP

Concentration cell An electrochemical cell composed of two identical electrodes immersed in two electrolytes that differ by their concentration

Concentration polarization Overvoltage contribution produced by concentration changes in the electrolyte

Corrosion Deterioration of a material, usually a metal, by a chemical or electrochemical reaction with its environment

Corrosion allowance Extra thickness which can be consumed by corrosion without affecting the integrity and resistance to pressure. It is obtained as corrosion rate multiplied by design life

Corrosion current The corrosion rate expressed in A/m^2 (for iron, 1 A/m^2 corresponds to 1.17 mm/year thickness loss)

Corrosion inhibitor A chemical substance that prevents or reduces the corrosion rate

Corrosion fatigue Fatigue-type cracking of metal caused by repeated or fluctuating stresses in a corrosive environment characterized by shorter life than would be encountered as a result of either the repeated or fluctuating stress alone or the corrosive environment alone

Corrosion potential (E_{corr}) The potential of a corroding surface in an electrolyte under open-circuit conditions (also known as rest potential, open-circuit potential, or free corrosion potential)

Corrosion rate The rate at which corrosion proceeds, usually expressed as μm/year

Corrosion resistance Ability of a material to withstand corrosion in a given system

Corrosiveness The tendency of an environment to cause corrosion

Crevice corrosion Localized corrosion of an active–passive metal at a shielded surface from full exposure to the environment (typically in a gap)

Current density The current to or from a unit area of an electrode surface usually expressed in mA/m^2

D

Depolarization The removal of overvoltage contributions

Differential aeration cell An electrochemical cell composed of two different compartments with electrodes of same metal and different oxygen concentration

Diffusion Displacement of atoms and ions under the effect of a concentration difference, governed by Fick Laws

Diffusion limiting current density The current density that corresponds to the maximum transfer rate that a particular species can sustain because of the limitation of diffusion (often referred to as limiting current density)

Double layer The interface between a metal and an electrolyte where an electrical charge separation takes place (the simplest model is represented by a parallel plate condenser)

Driving voltage The difference in potential which measures the metal's tendency to oxidize or more generally the energy available for a reaction to occur (also difference in potential between the structure and the anode)

E

Electrical continuity Ability of a structure composed of different materials to conduct electricity through electrical connections

Electrochemical cell A cell consisting of two electrodes, an anode and a cathode, immersed in an electrolyte

Electrochemical reaction A chemical reaction characterized by a gain or loss of electrons at electrode surfaces

Electrode A metal–electrolyte interface at which an electrochemical reaction proceeds or is in equilibrium

Electrode potential The potential of an electrode measured against a reference electrode

Electro-osmosis The phenomenon of water diffusion through a coating or a porous media promoted by the passage of current

Electrolysis The forced passage of electricity through a cell that produces chemical changes at the electrodes

Electrolyte A substance containing ions that migrate in an electric field

EMF series A list of elements arranged according to their standard electrode potentials. The sign is positive for elements whose potentials are more noble than hydrogen and negative for those less noble to hydrogen

Environment The surroundings or conditions (physical, chemical, mechanical) in which a material is exposed

Equilibrium potential The electrode potential of a reversible electrode when it is not polarized

Evans diagrams Diagram potential–log current density proposed by U.R. Evans (1889–1980) mostly used to represent corrosion processes

Erosion–corrosion A conjoint action involving corrosion and erosion in the presence of a moving corrosive fluid leading to accelerated loss of material

Exchange current density The rate of exchange of positive or negative charges between the metal–electrolyte interface of an electrode when the system is in equilibrium (dynamic equilibrium in an electrolyte)

Extreme value statistics It deals with the extreme deviations from the median of probability distributions. It aims to assess, from a given ordered sample of a given random variable, the probability of events that are more extreme than any previously observed

F

Faraday A quantity of electric charge, equal to 96,485 C, required to oxidize or reduce one chemical equivalent

Ferrite The body-centred cubic crystalline phase of iron-based alloys

Ferritic steel A steel whose microstructure at room temperature consists predominantly of ferrite

FEM Finite Element Method. A numerical technique for solving field equations based on imposing the constraints on anodic and cathodic surface and the discretization of the domain

Free corrosion potential E_{corr} The rest potential of a freely corroding metal in an electrolyte

Fick laws Field equations which describe diffusion phenomenon

Flade potential Defined by Friedrich Flade (1880–1916). See Passivity Potential (E_p)

Fouling Deposits including accumulation and growth of marine organisms, beneath which biofilm can form

G

Galvanic anodes Metals and alloys used to obtain CP of more noble materials

Galvanic cell An electrochemical cell with electrode reaction proceeding spontaneously and liberating energy

Galvanic corrosion Accelerated corrosion of a metal because of an electrical contact with a more noble material (a metal or a nonmetallic conductor) in a corrosive electrolyte (also called bimetallic corrosion)

Galvanic coupling A pair of dissimilar metals or conductive nonmetals in electrical contact in an electrolyte (also called bimetallic coupling)

Galvanic current The electric current in a galvanic cell

Galvanic series Ranking of metals and alloys according to their corrosion potentials in a given electrolyte

Galvanostatic Refers to a technique for maintaining a constant exchanged current on an electrode

General corrosion Corrosion that is distributed more or less uniformly over the surface of a material

Gumbel distribution It is one of most used Extreme Value Statistics, often used for localized corrosion analysis (for instance, maximum pit depth)

H

HSLA (High Strength Low-Alloy Steel) Low-alloy steels that are heat treated to reach high tensile resistance

Hydrogen embrittlement Loss of ductility of a metal resulting from the absorption of atomic hydrogen produced in the hydrogen evolution cathodic reaction

Hydrogen-induced cracking (HIC) Internal damage as small cracks (stepwise type) caused by atomic hydrogen on carbon steel with high sulphur content

Hydrogen overvoltage Overvoltage associated with the reaction of hydrogen evolution

I

Immunity A metal's state of corrosion resistance that corresponds to an electrode potential more negative than the equilibrium potential

Imperfect passivity A passivity condition of an active–passive metal which resists to localized corrosion (pitting and crevice) if never started but cannot stop it if initiated

Impressed current anode An anode used in impressed CP systems. It stands an anodic reaction as metal dissolution, oxygen or chlorine evolution

Impressed current An electric current supplied by a device employing a power source that is external to the electrode system

Impressed current CP CP obtained by impressed current through an impressed current anode

Intensification index It is used in corrosion testing as ratio between the corrosion rate in accelerated tests and the expected corrosion rate in operating plant

Intergranular corrosion Type of localized corrosion attack occurring at grain boundaries of sensitized alloys, typically stainless steels and nickel alloys

Ion An electrically charged atom or complex of atoms

Ion activity The molar concentration of an ion multiplied by the mean ion-activity coefficient

IR drop The voltage that results from the flow of current through a conductor (metallic or electrolytic)

L

Langelier Saturation Index (LSI) The Langelier Saturation index (LSI) is an indicator of the degree of saturation of calcium carbonate in water. It is calculated as $pH - pH_S$ (actual pH minus saturation pH of calcium carbonate). Positive values indicate tendency to form calcium carbonate scales

Localized corrosion Corrosion attack occurring in a small portion of the exposed surface of metal

Low-alloy steel Steel containing intentional alloying element additions with a content less than 5%

LPR (Linear Polarization Resistance) Electrochemical technique used to measure the instantaneous corrosion rate of a metal in an electrolyte. It employs three electrodes: a working electrode, a reference electrode and an auxiliary electrode (or counter electrode)

Lugging probe A small tube or a capillary filled with electrolyte, terminating closely to the metal surface of an electrode to eliminate ohmic drop during potential measurements

M

Macrocell Electrochemical cell setting up by the separation of anodic and cathodic surfaces

Martensitic stainless steel A stainless steel with martensitic microstructure at room temperature

Median rank Equation proposed by Fisher (1890–1962) for converting a series of data of a sample into an approximating cumulative distribution function

MIC Microbiologically Influenced Corrosion (MIC), also known as microbial corrosion or biological corrosion, is a corrosion attack promoted by the presence of specific bacteria

MOB Manganese-Oxidizing Bacteria (MOB) are bacteria of Leptothrix strain type which ennobles the redox potential, then inducing pitting attack on low-grade stainless steels in freshwater

N

Nernst equation An equation that expresses the equilibrium potential of an electrochemical reaction

Noble Expresses a low tendency to react or the positive direction of electrode potential

Noble metal Metal with a low tendency to react (or corrode), such as gold or platinum. Its potential is highly positive

Noble potential A potential more cathodic (positive) than the standard hydrogen potential

O

Open-circuit potential The potential of an electrode in the absence of an external current flow

Overvoltage The change in potential of an electrode from its equilibrium when current is exchanged with the electrolyte (see Polarization)

Oxidation (1) An electrochemical reaction in which electrons are released. (2) Corrosion of a metal that is exposed to an oxidizing gas typically at elevated temperatures

Oxidation–reduction potential The potential of a reversible oxidation–reduction electrode in a given electrolyte

Oxygen limiting current density Maximum quantity of oxygen that can reach the metal surface through diffusion; it can be calculated by Fick Laws

P

Passivation A reduction of the anodic reaction rate of a metal by the formation of a protective oxide film

Passivation potential The potential that an active–passive metal shows when it is in passive state

Passive A state of passivation of a metal

Passivity The state of being passive. It is a condition in which an active–passive metal has a potential much more noble than the active state

Passivity potential (E_p) Also called Flade potential. It is the lower value of the passive interval

Pedeferri diagram (E-%Cl) A diagram potential–chloride content proposed by Pietro Pedeferri (1938–2008) to represent graphically the regions for occurrence or prevention of pitting of passive metals

Perfect passivity A passivity condition of a metal which resists to localized corrosion (pitting and crevice)

pH A measure of the acidity or alkalinity of an aqueous electrolyte. pH value is the negative logarithm of the hydrogen ion activity ($-\log[H^+]$ where $[H^+]$ = hydrogen ion activity)

Pitting Localized corrosion of an active–passive metal that takes the form of cavities called pits. Surface area surrounding pits remain passive. Pitting is an example of infant mortality corrosion attack

Pitting equation and Pitting induction time PIT is period of time for having pitting initiation which corresponds to the passive film breakdown and the corrosion macrocell sets up

PIN Pitting initiation number is an index of stainless steels which is calculated from their composition and takes into account the presence of inclusions

Pitting potential (E_{pit}) The lowest value of potential at which pits nucleate and grow

Polarization The change of potential caused by a current flow across the electrode/electrolyte interface

Polarization curve A plot of current density versus electrode potential for a specific electrode/electrolyte combination

Polarization resistance The slope (dE/di) at the corrosion potential of a potential (E)–current density (i) curve. Corrosion rate is inversely proportional to the polarization resistance (see Linear Polarization Resistance)

Porosity Ratio of covered surface area of a metal by a coating/layer to the remaining bare metal surface. Complement of coating efficiency

Potential-pH diagram (Pourbaix diagram) A graphical representation of thermodynamic stability of species for metal/electrolyte systems proposed by M. Pourbaix (1904–1998)

Potentiodynamic It refers to a technique wherein the potential of an electrode with respect to a reference electrode is varied at a selected rate by application of an external current

Potentiostat An instrument for automatically maintaining a constant electrode potential

Potentiostatic It refers to a technique for maintaining a constant electrode potential

PREN Pitting Resistance Evaluation Number is an index of stainless steels which is calculated from their composition

Protection current density The current density necessary to obtain CP of a metal in an environment. It equals the rate of cathodic processes taking place at the protection potential

Primary distribution Current distribution on an electrode obtained when overvoltage (polarization) is nil or negligible. It depends on geometry of the domain (anode and cathode) only and not on electrolyte resistivity

Pseudo-reference electrode An electrode with no thermodynamic potential but sufficiently stable to be used as a reference electrode for instant measurements

Q

Quasi-immunity A state of metal when its potential is slightly more positive than the equilibrium potential so that corrosion rate is negligible

R

Redox potential (see Oxidation–Reduction Potential)

Reduction An electrochemical reaction in which electrons are acquired

Reference electrode An electrode with stable potential used as a reference to measure the potential of another electrode or a structure under CP

Relative humidity The ratio, expressed as a percentage, of the amount of water vapor present in a given volume of air at a given temperature to the amount required to saturate the air at that temperature

Rest potential Potential measured on an electrode through which no net current is flowing (see open-circuit potential)

Rust Corrosion product of iron and steel consisting of various iron oxides and hydrated iron oxides

S

Sampling Procedure to minimize bias errors in selecting elements from a population for testing. Types of sampling are random, systematic, stratified, cluster

Scaling (1) The formation at high temperatures of thick corrosion-product layers on a metal surface. (2) The deposition of water-insoluble constituents on a metal surface

Secondary distribution Current distribution on an electrode obtained when overvoltage takes place. It depends on both geometry of the domain (anode and cathode) and electrolyte resistivity

Sensitization (sensitized alloy) Sensitization of a stainless steel and nickel alloys is the presence of a chromium-depleted zone around the chromium carbides precipitated at grain boundaries. It happens typically during welding at 425–815°C temperature range

Solubilization treatment Heat treatment of stainless steels and nickel alloys aiming to dissociate carbides and dissolve carbon in the metal matrix. Often it consists of heating at 1050°C for 1 h per 1-in. thickness followed by a rapid cooling

Stabilized stainless steels Addition of stabilizers such as niobium or titanium avoids the chromium carbides precipitation during welding, then impeding sensitization

SRB Sulphate-Reducing Bacteria (SRB) are anaerobic bacteria, thriving in anoxic environment whilst not dying in the presence of oxygen. They catalyse the reduction of sulphate to sulphide then leading to corrosion conditions of iron

Stainless steels Ferrous alloys with a minimum chromium content of 12 wt% to obtain passivation of the surface

Standard electrode potential The reversible potential for an electrode process when all products and reactions are at unitary activity on a scale in which the potential for the standard hydrogen reference electrode is zero

Stress corrosion cracking (SCC) Cracking of a material produced by the combined action of corrosion and tensile stress (residual or applied)

Sulphide stress corrosion cracking (SSC) Form of stress corrosion cracking (SCC) which proceeds by hydrogen embrittlement mechanism in the presence of hydrogen sulphide

T

Tafel plot A plot of the relationship between the change in potential (E) and the logarithm of the current density (log i) of an electrode when it is polarized in both the anodic and cathodic directions from its open-circuit potential

Tafel slope The slope of the straight-line portion of the E–log i curve on a Tafel plot. (The straight-line portion usually occurs at more than 50 mV from the open-circuit potential)

TDS Total dissolved solid represents the amount of salts dissolved in a water

Throwing Power It is the distance at which macrocell current vanishes. In CP, it is the distance from the anode of the point of the cathodic surface still in protection

True potential The potential of an electrode (a structure) free from ohmic drop contributions. It is the sum of the equilibrium potential and overvoltage. It can be measured by means of potential probes, Lugging capillary or on–off technique (also called off-potential)

UV

Valence Number of electrons released or gained by an atom or a molecule

WXYZ

Wetting factor Fraction of time in which an electrolyte is in contact with a metal surface

Working electrode The test or specimen electrode in an electrochemical cell submitted to a finite current density

Weibull distribution It is one of most used extreme value statistics, often used for localized corrosion analysis (for instance, lowest time-to-failure in SCC)

Index

Note: Page numbers followed by "*f*" and "*t*" refer to figures and tables, respectively.

T

FSC
www.fsc.org
MIX
Papier | Fördert gute Waldnutzung
FSC® C083411

Kolibri 360 GmbH
Ferdinand-Jühlke-Straße 7
99095 Erfurt, Deutschland
produktsicherheit@kolibri360.de